DATA STRUCTURES
AND
PROGRAM DESIGN
USING JAVA

DATA STRUCTURES

AND

PROGRAM DESIGN

USING JAVA

A Self-Teaching Introduction

Dheeraj Malhotra, PhD
Neha Malhotra, PhD

MERCURY LEARNING AND INFORMATION

Dulles, Virginia
Boston, Massachusetts
New Delhi

Publisher: David Pallai
MERCURY LEARNING AND INFORMATION
22841 Quicksilver Drive
Dulles, VA 20166
info@merclearning.com
www.merclearning.com
(800) 232-0223

D. Malhotra and N. Malhotra. *Data Structures and Program Design Using JAVA.*
ISBN: 978-1-68392-464-7

The publisher recognizes and respects all marks used by companies, manufacturers, and developers as a means to distinguish their products. All brand names and product names mentioned in this book are trademarks or service marks of their respective companies. Any omission or misuse (of any kind) of service marks or trademarks, etc. is not an attempt to infringe on the property of others.

Library of Congress Control Number: 2020930984

202122321   Printed on acid-free paper in the United States of America.

Our titles are available for adoption, license, or bulk purchase by institutions, corporations, etc. For additional information, please contact the Customer Service Dept. at (800) 232-0223 (toll free). Digital versions of our titles are available at: *www.academiccourseware.com* and other electronic vendors.

Dedicated to our
loving parents and beloved students

CONTENTS

Preface   *xiii*

Acknowledgments   *xv*

Chapter 1   Introduction to Data Structures   **1**

1.1   Introduction   1

1.2   Types of Data Structures   2

   1.2.1   Linear and Non-Linear Data Structures   3

   1.2.2   Static and Dynamic Data Structures   3

   1.2.3   Homogeneous and Non-Homogeneous Data Structures   3

   1.2.4   Primitive and Non-Primitive Data Structures   3

   1.2.5   Arrays   4

   1.2.6   Queues   5

   1.2.7   Stacks   6

   1.2.8   Linked Lists   6

   1.2.9   Trees   8

   1.2.10   Graphs   9

1.3   Operations on Data Structures   10

1.4   Algorithms   10

   1.4.1   Developing an Algorithm   11

1.5   Approaches for Designing an Algorithm   12

1.6   Analyzing an Algorithm   13

   1.6.1   Time-Space Trade-Off   14

1.7   Abstract Data Types   14

1.8   Big O Notation   15

1.9   Summary   16

1.10   Exercises   17

   1.10.1   Theory Questions   17

   1.10.2   Multiple Choice Questions   18

Chapter 2   Introduction to the Java Language   **21**

2.1   Introduction   21

2.2   Java and Its Characteristics   22

2.3   Java Overview   22

2.4   Compiling the Java Program   23

2.5   Object-Oriented Programming   24

2.6   Character Set Used in Java   28

2.7   Java Tokens   29

2.8   Data Types in Java   30

2.9   Structure of a Java Program   31

2.10   Operators in Java   32

2.11   Decision Control Statements in Java   36
2.12   Looping Statements in Java   45
2.13   Break and Continue Statements   51
2.14   Methods in Java   54
2.15   Summary   55
2.16   Exercises   57
    2.16.1   Theory Questions   57
    2.16.2   Programming Questions   58
    2.16.3   Multiple Choice Questions   59

Chapter 3   Arrays   61
3.1   Introduction   61
3.2   Definition of an Array   61
3.3   Array Declaration   62
3.4   Array Initialization   63
3.5   Calculating the Address of Array Elements   64
3.6   Operations on Arrays   65
3.7   2-D Arrays/Two-Dimensional Arrays   83
3.8   Declaration of Two-Dimensional Arrays   84
3.9   Operations on 2-D Arrays   86
3.10   Multidimensional Arrays/N-Dimensional Arrays   90
3.11   Calculating the Address of 3-D Arrays   91
3.12   Arrays and Their Applications   93
3.13   Sparse Matrices   93
3.14   Types of Sparse Matrices   94
3.15   Representation of Sparse Matrices   95
3.16   Summary   96
3.17   Exercises   98
    3.17.1   Theory Questions   98
    3.17.2   Programming Questions   99
    3.17.3   Multiple Choice Questions   99

Chapter 4   Linked Lists   103
4.1   Introduction   103
4.2   Definition of a Linked List   103
4.3   Memory Allocation in a Linked List   105
4.4   Types of Linked Lists   106
    4.4.1   Singly Linked List   106
    4.4.2   Operations on a Singly Linked List   106
    4.4.3   Circular Linked Lists   122
    4.4.4   Operations on a Circular Linked List   123
    4.4.5   Doubly Linked List   133
    4.4.6   Operations on a Doubly Linked List   134

4.5	Header Linked Lists	149
4.6	Applications of Linked Lists	159
4.7	Polynomial Representation	159
4.8	Summary	159
4.9	Exercises	160
4.9.1	Theory Questions	160
4.9.2	Programming Questions	160
4.9.3	Multiple Choice Questions	161

Chapter 5   Queues   **163**
5.1	Introduction	163
5.2	Definition of a Queue	163
5.3	Implementation of a Queue	164
5.3.1	Implementation of Queues Using Arrays	164
5.3.2	Implementation of Queues Using Linked Lists	164
5.3.2.1	Insertion in Linked Queues	165
5.3.2.2	Deletion in Linked Queues	166
5.4	Operations on Queues	170
5.4.1	Insertion	170
5.4.2	Deletion	171
5.5	Types of Queues	175
5.5.1	Circular Queue	175
5.5.1.1	Limitation of Linear Queues	176
5.5.1.2	Inserting an Element in a Circular Queue	178
5.5.1.3	Deleting an Element from a Circular Queue	180
5.5.2	Priority Queue	185
5.5.2.1	Implementation of a Priority Queue	186
5.5.2.2	Insertion in a Linked Priority Queue	188
5.5.2.3	Deletion in a Linked Priority Queue	188
5.5.3	De-Queues (Double-Ended Queues)	192
5.6	Applications of Queues	197
5.7	Summary	197
5.8	Exercises	198
5.8.1	Theory Questions	198
5.8.2	Programming Questions	198
5.8.3	Multiple Choice Questions	199

Chapter 6   Searching and Sorting   **201**
6.1	Introduction to Searching	201
6.2	Linear Search or Sequential Search	201
6.2.1	Drawbacks of a Linear Search	204
6.3	Binary Search	206
6.3.1	Binary Search Algorithm	206

6.3.2   Complexity of a Binary Search Algorithm   208
6.3.3   Drawbacks of a Binary Search   208
6.4   Interpolation Search   210
6.4.1   Working of the Interpolation Search Algorithm   211
6.4.2   Complexity of the Interpolation Search Algorithm   212
6.5   Introduction to Sorting   214
6.5.1   Types of Sorting Methods   215
6.6   External Sorting   235
6.7   Summary   236
6.8   Exercises   237
6.8.1   Theory Questions   237
6.8.2   Programming Questions   237
6.8.3   Multiple Choice Questions   238

Chapter 7   Stacks   **241**
7.1   Introduction   241
7.2   Definition of a Stack   242
7.3   Overflow and Underflow in Stacks   242
7.4   Operations on Stacks   243
7.5   Implementation of Stacks   249
7.5.1   Implementation of Stacks Using Arrays   249
7.5.2   Implementation of Stacks Using Linked Lists   249
7.5.2.1   Push Operation in Linked Stacks   250
7.5.2.2   Pop Operation in Linked Stacks   251
7.6   Applications of Stacks   255
7.6.1   Polish and Reverse Polish Notations   255
7.6.2   Conversion from Infix Expression to Postfix Expression 256
7.6.3   Conversion from Infix Expression to Prefix Expression 262
7.6.4   Evaluation of a Postfix Expression   266
7.6.5   Evaluation of a Prefix Expression   270
7.6.6   Parenthesis Balancing   274
7.7   Summary   278
7.8   Exercises   278
7.8.1   Theory Questions   278
7.8.2   Programming Questions   280
7.8.3   Multiple Choice Questions   280

Chapter 8   Trees   **283**
8.1   Introduction   283
8.2   Definitions   284
8.3   Binary Tree   287
8.3.1   Types of Binary Trees   288
8.3.2   Memory Representation of Binary Trees   289

8.4   Binary Search Tree   291
    8.4.1   Operations on Binary Search Trees   291
    8.4.2   Binary Tree Traversal Methods   305
    8.4.3   Creating a Binary Tree Using Traversal Methods   315
8.5   AVL Trees   319
    8.5.1   Need of Height-Balanced Trees   319
    8.5.2   Operations on an AVL Tree   320
8.6   Summary   330
8.7   Exercises   332
    8.7.1   Theory Questions   332
    8.7.2   Programming Questions   334
    8.7.3   Multiple Choice Questions   335

Chapter 9   Multi-Way Search Trees   339
9.1   Introduction   339
9.2   B-Trees   340
9.3   Operations on a B-Tree   341
    9.3.1   Insertion in a B-Tree   341
    9.3.2   Deletion in a B-Tree   343
9.4   Application of a B-Tree   349
9.5   B+ Trees   349
9.6   Summary   350
9.7   Exercises   351
    9.7.1   Review Questions   351
    9.7.2   Multiple Choice Questions   351

Chapter 10   Hashing   353
10.1   Introduction   353
    10.1.1   Difference between Hashing and Direct Addressing   354
    10.1.2   Hash Tables   355
    10.1.3   Hash Functions   356
    10.1.4   Collision   358
    10.1.5   Collision Resolution Techniques   358
    10.1.5.1   Chaining Method   358
    10.1.5.2   Open Addressing Method   363
10.2   Summary   376
10.3   Exercises   378
    10.3.1   Review Questions   378
    10.3.2   Multiple Choice Questions   379

Chapter 11   Files   381
11.1   Introduction   381
11.2   Terminologies   381
11.3   File Operations   382

11.4   File Classification   383
11.5   C vs C++ vs Java File Handling   384
11.6   File Organization   384
11.7   Sequence File Organization   385
11.8   Indexed Sequential File Organization   386
11.9   Relative File Organization   387
11.10   Inverted File Organization   388
11.11   Summary   388
11.12   Exercises   389
  11.12.1   Review Questions   389
  11.12.2   Multiple Choice Questions   390

Chapter 12   Graphs   **393**
12.1   Introduction   393
12.2   Definitions   394
12.3   Graph Representation   398
  12.3.1   Adjacency Matrix Representation   398
  12.3.2   Adjacency List Representation   400
12.4   Graph Traversal Techniques   402
  12.4.1   Breadth First Search   402
  12.4.2   Depth First Search   406
12.5   Topological Sort   410
12.6   Minimum Spanning Tree   414
  12.6.1   Prim's Algorithm   414
  12.6.2   Kruskal's Algorithm   416
12.7   Summary   418
12.8   Exercises   419
  12.8.1   Theory Questions   419
  12.8.2   Programming Questions   421
  12.8.3   Multiple Choice Questions   422

Answers to Multiple Choice Questions   **425**

Index   *427*

PREFACE

Data structures are the building blocks of computer science. The objective of this text is to emphasize the fundamentals of data structures as an introductory subject. It is designed for beginners who would like to learn the basics of data structures and their implementation using the Java programming language. With this focus in mind, we present various fundamentals of the subject, well supported with real-world analogies to enable a quick understanding of the technical concepts and to help in identifying appropriate data structures to solve specific, practical problems. This book will serve the purpose of a text/reference book and will be of immense help especially to undergraduate or graduate students of various courses in information technology, engineering, computer applications, and information sciences.

Key Features

- *Practical Applications:* Real world analogies as practical applications are given throughout the text to quickly understand and connect the fundamentals of data structures with day to day, real-world scenarios. This approach, in turn, will assist the reader in developing the capability to identify the most appropriate and efficient data structure for solving a specific, real-world problem.

- *Frequently Asked Questions:* Frequently asked theoretical/practical questions are integrated throughout the content of the book, within related topics to assist readers in grasping the subject.

- *Algorithms and Programs:* To better understand the fundamentals of data structures at a generic level-followed by its object-oriented implementation in Java, syntax independent algorithms, as well as implemented programs in Java, are discussed throughout the book. This presentation will assist the reader in getting both algorithms and their corresponding implementation within a single book.

- *Numerical and Conceptual Exercises:* To assist the reader in developing a strong foundation of the subject, various numerical and conceptual problems are included throughout the text.

- *Multiple Choice Questions:* To assist students for placement-oriented exams in various IT fields, several exercises are suitably chosen and are given in an MCQ format.

Dr. Dheeraj Malhotra
Dr. Neha Malhotra
February 2020

ACKNOWLEDGMENTS

We are indeed grateful to Chairman VIPS- Dr. S.C. Vats, respected management, Chairperson VSIT, and Dean VSIT of Vivekananda Institute of Professional Studies (GGS IP University). They are always a source of inspiration for us, and we feel honored because of their faith in us.

We also take this opportunity to extend our gratitude to our mentors Dr. O.P. Rishi (University of Kota), Dr. Sushil Chandra (DRDO, GOI), and Dr. Udyan Ghose (GGS IP University) for their motivation to execute this project.

We are profoundly thankful to Ms. Stuti Suthar (VIPS, GGSIPU) and Mr. Deepanshu Gupta (Tech Mahindra Ltd.) for helping us in proofreading and compiling the codes in this manuscript.

It is not possible to complete a book without the support of a publisher. We are thankful to David Pallai and Jennifer Blaney of Mercury Learning and Information for their enthusiastic involvement throughout the tenure of this project.

Our heartfelt regards to our parents, siblings and family members who cheered us in good times and encouraged us in bad times.

Lastly, we have always felt inspired by our readers especially in the USA, Canada, and India. Their utmost love and positive feedback for our first two titles of *Data Structures using C* and *C++*, both published with MLI, helped us to further improve the current title.

Dr. Dheeraj Malhotra
Dr. Neha Malhotra
February 2020

CHAPTER 1

*I*NTRODUCTION TO *D*ATA *S*TRUCTURES

1.1 Introduction

A data structure is an efficient way of storing and organizing data elements in the computer memory. Data means a value or a collection of values. Structure refers to a method of organizing the data. The mathematical or logical representation of data in the memory is referred as a data structure. The objective of a data structure is to store, retrieve, and update the data efficiently. A data structure can be referred to as elements grouped under one name. The data elements are called members, and they can be of different types. Data structures are used in almost every program and software system. There are various kinds of data structures that are suited for different types of applications. Data structures are the building blocks of a program. For a program to run efficiently, a programmer must choose appropriate data structures. A data structure is a crucial part of data management. As the name suggests, data management is a task which includes different activities like the collection of data, the organization of data into structures, and much more. Some examples where data structures are usedinclude stacks, queues, arrays, binary trees, linked lists, hash tables, and so forth.

A data structure helps usto understand the relationship of one element to another element and organize it within the memory. It is a mathematical or logical representation or organization of data in memory. Data structures are extensively applied in the following areas:

- Compiler Design
- Database Management Systems (DBMS)
- Artificial Intelligence
- Network and Numerical Analysis
- Statistical Analysis Packages
- Graphics
- Operating Systems (OS)
- Simulations

As we see in the previous list, there are many applications in which different data structures are used for their operations. Some data structures sacrifice speed for efficient utilization of memory, while others sacrifice memory utilization and result in faster speed. In today's world programmers aim not just to build a program but instead to build an effective program. As previously discussed, for a program to be efficient, a programmer must choose appropriate data structures. Hence, data structures are classified into various types. Now, let us discuss and learn about different types of data structures.

Frequently Asked Questions
1. Define the term data structure.
Ans: *A data structure is an organization of data in a computer's memory or disk storage. In other words, a logical or mathematical model of a particular organization of data is called a data structure. A data structure in computer science is also a way of storing data in a computer so that it can be used efficiently. An appropriate data structure allows a variety of important operations to be performed using both resources, that is, memory space and execution time, efficiently.*

1.2 Types of Data Structures

Data structures are classified into various types.

1.2.1 Linear and Non-Linear Data Structures

A linear data structure is one in which the data elements are stored in a linear, or sequential, order; that is, data is stored in consecutive memory locations. A linear data structure can be represented in two ways; either it is represented by a linear relationship between various elements utilizing consecutive memory locations as in the case of arrays, or it may be represented by a linear relationship between the elements utilizing links from one element to another as in the case of linked lists. Examples of linear data structures include arrays, linked lists, stacks, queues, and so on.

A non-linear data structure is one in which the data is not stored in any sequential order or consecutive memory locations. The data elements in this structure are represented by a hierarchical order. Examples of non-linear data structures include graphs, trees, and so forth.

1.2.2 Static and Dynamic Data Structures

A static data structure is a collection of data in memory which is fixed in size and cannot be changed during runtime. The memory size must be known in advance, as the memory cannot be reallocated later in a program. One example is an *array*.

A dynamic data structure is a collection of data in which memory can be reallocated during the execution of a program. The programmer can add or remove elements according to his or her need. Examples include linked lists, graphs, trees, and so on.

1.2.3 Homogeneous and Non-Homogeneous Data Structures

A homogeneous data structure is one that contains data elements of the same type, for example, arrays.

A non-homogeneous data structure contains data elements of different types, for example, structures.

1.2.4 Primitive and Non-Primitive Data Structures

Primitive data structures are fundamental data structures or predefined data structures which are supported by a programming language. Examples of primitive data structure types are integer, float, char, and so forth.

Non-primitive data structures are comparatively more complicated data structures that are created using primitive data structures. Examples of non-primitive data structures are arrays, files, linked lists, stacks, queues, and so on.

The classification of different data structures is shown in Figure 1.1.

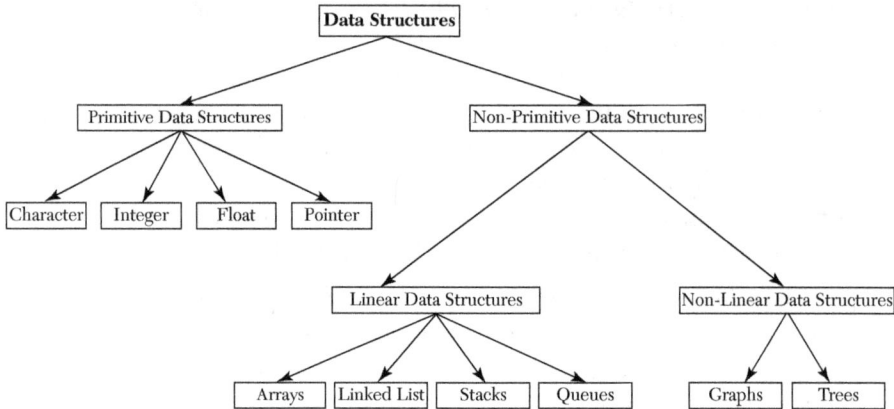

Figure 1.1. Classification of different data structures.

We know that Java supports various data structures. So, we will now introduce all these data structures, and they will be discussed in detail in the upcoming chapters.

Frequently Asked Questions

2. Write the difference between primitive data structures and non-primitive data structures.

Ans: *Primitive data structures: The data structures that are typically directly operated upon by machine-level instructions, that is, the fundamental data types such as int, float, char, and so on, are known as primitive data structures.*

   Non-primitive data structures: The data structures which are not fundamental are called non-primitive data structures.

3. Explain the difference between linear and non-linear data structures.

Ans: *The main difference between linear and non-linear data structures lies in the way in which data elements are organized. In a linear data structure, elements are organized sequentially, and therefore they are easy to implement in a computer's memory. In non-linear data structures, a data element can be attached to several other data elements to represent specific relationships existing among them.*

1.2.5   Arrays

An array is a collection of homogeneous (similar) types of data elements in contiguous memory. An array is a linear data structure, because all elements of an array are stored in linear order. The various elements of the array are referenced by their index value, also known as the subscript. In Java, an array is declared using the following syntax:

```
Syntax – <Data type> array name [size];
```

The elements are stored in the array as shown in Figure 1.2.

1st element	2nd element	3rd element	4th element	5th element	6th element
array[0]	array[1]	array[2]	array[3]	array[4]	array[5]

Figure 1.2. Memory representation of an array.

Arrays are used for storing a large amount of data of similar type. They have various advantages and limitations.

Advantages of using arrays

1. Elements are stored in adjacent memory locations; hence, searching is very fast, as any element can be easily accessed.

2. Arrays do not support dynamic memory allocation, so all the memory management is done by the compiler.

Limitations of using arrays

1. Insertion and deletion of elements in arrays is complicated and very time-consuming, as it requires the shifting of elements.

2. Arrays are static; hence, the size must be known in advance.

3. Elements in the array are stored in consecutive memory locations which may or may not be available.

1.2.6 Queues

*A queue is a linear collection of data elements*in which the element inserted first will be the element that is taken out first; that is, a queue is a FIFO (First In First Out) data structure. A queue is a popular linear data structure in which the first element is inserted from one end called the REAR end (also called the tail end), and the deletion of the element takes place from the other end called the FRONT end (also called the head).

Practical Application

For a simple illustration of a queue, there is a line of people standing at the bus stop and waiting for the bus. Therefore, the first person standing in the line will get into the bus first.

In a computer's memory queues can be implemented using arrays or linked lists. Figure 1.3 shows the array implementation of a queue. Every queue has FRONT and REAR variables which point to the positions where deletion and insertion are done respectively.

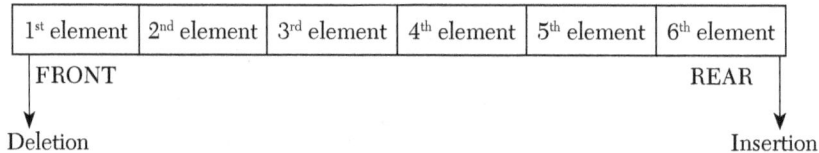

1st element	2nd element	3rd element	4th element	5th element	6th element

FRONT   REAR

Deletion   Insertion

Figure 1.3. Memory representation of a queue.

1.2.7 Stacks

A stack is a linear collection of data elements in which insertion and deletion take place only at the top of the stack. A stack is a Last In First Out (LIFO) data structure, because the last element pushed onto the stack will be the first element to be deleted from the stack. The three operations that can be performed on the stack include the PUSH, POP, and PEEP operations. The PUSH operation inputs an element into the top of the stack, while the POP operation removes an element from the stack. The PEEP operation returns the value of the topmost element in the stack without deleting it from the stack. Every stack has a variable TOP which is associated with it. The TOP node stores the address of the topmost element in the stack. The TOP is the position where insertion and deletion take place.

Practical Application

A real-life example of a stack is if there is a pile of plates arranged on a table. A person will pick up the first plate from the top of the stack.

In a computer's memory stacks can be implemented using arrays or linked lists. Figure 1.4 shows the array implementation of a stack.

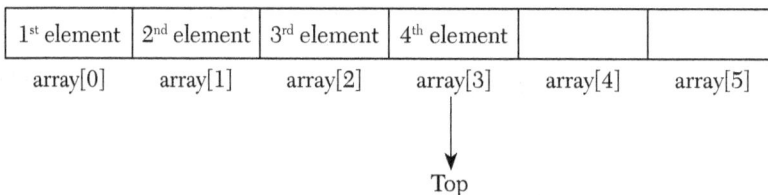

1st element	2nd element	3rd element	4th element		
array[0]	array[1]	array[2]	array[3]	array[4]	array[5]

Top

Figure 1.4. Memory representation of a stack.

1.2.8 Linked Lists

The major drawback of the array is that the size or the number of elements must be known in advance. Thus, this drawback gave rise to the new concept of a linked list. *A linked list is a linear collection of data elements.* These data elements are called nodes, and each node stores the address of the next node. *A linked list is a sequence of nodes in which each node contains one or more than one data field and an address field that stores the address of the next node.* Also, linked lists are dynamic; that is, memory is allocated as and when required.

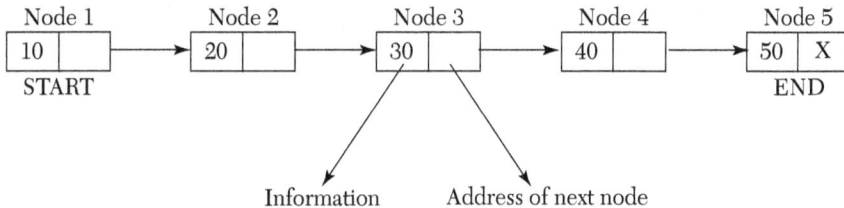

Figure 1.5. Memory representation of a linked list.

In the previous figure, we have made a linked list in which each node is divided into two slots:

1. The first slot contains the information/data.

2. The second slot contains the address of the next node.

Practical Application

A simple real-life example is a train; here each coach is connected to its previous and next coach (except the first and last coach).

The address part of the last node stores a special value called NULL, which denotes the end of the linked list. The advantage of a linked list over arrays is that now it is easier to insert and delete data elements, as we don't have to do shifting each time. Yet searching for an element has become difficult. Also, more time is required to search for an element, and it also requires high memory space. Hence, linked lists are used where a collection of data elements is required but the number of data elements in the collection is not known to us in advance.

Frequently Asked Questions

4. Define the term linked list.

Ans: *A linked list or one-way list is a linear collection of data elements called nodes, which give a linear order. It is a popular dynamic data structure. The nodes in the linked list are not stored in consecutive memory locations. For every data item in a node of the linked list, there is an associated address field that gives the address location of the next node in the linked list.*

1.2.9 Trees

A tree is a popular non-linear data structure in which the data elements or the nodes are represented in a hierarchical order. Here, one of the nodes is shown as the root node of the tree, and the remaining nodes arepartitioned into two disjointed sets such that each set is a part of a subtree. A tree makes the searching process very easy, and its recursive programming makes a program optimized and easy to understand.

A binary tree is the simplest form of a tree. *A binary tree consists of a root node and two subtrees known as the left subtree and the right subtree, where both subtrees are also binary trees.* Each node in a tree consists of three parts, that is, the extreme left part stores the address of the left subtree, the middle part consists of the data element, and the extreme right part stores the address of the right subtree. The root is the topmost element of the tree. When there are no nodes in a tree, that is, when ROOT = NULL, then it is called an empty tree.

For example, consider a binary tree where R is the root node of the tree. LEFT and RIGHT are the left and right subtrees of R respectively. Node A is designated as the root node of the tree. Nodes B and C are the left and right child of A respectively. Nodes B, D, E, and G constitute the left subtree of the root. Similarly, nodes C, F, H, and I constitute the right subtree of the root.

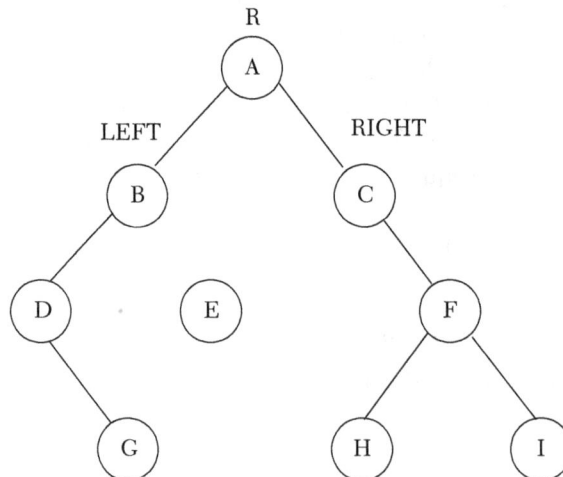

Figure 1.6. A binary tree.

Advantages of a tree

1. The searching process is very fast in trees.

2. Insertion and deletion of the elements have become easier as compared to other data structures.

Frequently Asked Questions

5. Define the term binary tree.

Ans: *A binary tree is a hierarchal data structure in which each node has at most two children, that is, a left and right child. In a binary tree, the degree of each node can be at most two. Binary trees are used to implement binary search trees, which are used for efficient searching and sorting. A variation of BST is an AVL tree where the height of the left and right subtree differs by one. A binary tree is a popular subtype of a k-ary tree, where k is 2.*

1.2.10   Graphs

A graph is a general tree with no parent-child relationship. It is a non-linear data structure which consists of vertices, also called nodes, and edges which connect those vertices to one another. In a graph, any complex relationship can exist. A graph G may be defined as a finite set of V vertices and E edges. Therefore, G = (V, E) where V is the set of vertices and E is the set of edges. Graphs are used in various applications of mathematics and computer science. Unlike a root node in trees, graphs don't have root nodes; rather, the nodes can be connected to any node in the graph. Two nodes are termed as neighbors when they are connected via an edge.

Practical Application

A real-life example of a graph can be seen in workstations where several computers are joined to one another via network connections.

For example, consider a graph G with six vertices and eight edges. Here, Q and Z are neighbors of P. Similarly, R and T are neighbors of S.

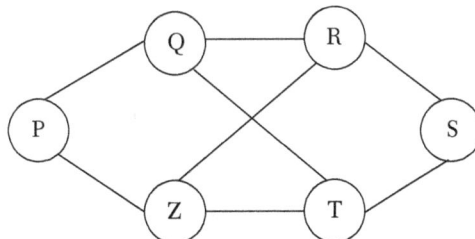

Figure 1.7. A graph.

1.3   Operations on Data Structures

Here we will discuss various operations which are performed on data structures.

- **Creation:** It is the process of creating a data structure. Declaration and initialization of the data structure are done here. It is the first operation.

- **Insertion:** It is the process of adding new data elements in the data structure, for example, to add the details of an employee who has recently joined an organization.

- **Deletion:** It is the process of removing a particular data element from the given collection of data elements, for example, to remove the name of an employee who has left the company.

- **Updating:** It is the process of modifying the data elements of a data structure. For example, if the address of a student is changed, then it should be updated.

- **Searching:** It is used to find the location of a particular data element or all the data elements with the help of a given key, for example, to find the names of people who live in New York.

- **Sorting:** It is the process of arranging the data elements in some order, that is, either an ascending or descending order. An example is arranging the names of students of a class in alphabetical order.

- **Merging:** It is the process of combining the data elements of two different lists to form a single list of data elements.

- **Traversal:** It is the process of accessing each data element exactly once so that it can be processed. An example is to print the names of all the students of a class.

- **Destruction:** It is the process of deleting the entire data structure. It is the last operation in the data structure.

1.4   Algorithms

An algorithm is a systematic set of instructions combined to solve a complex problem. It is a step-by-finite-step sequence of instructions, each of which has a clear meaning and can be executed with a minimum amount of effort in finite time. In general, an algorithm is a blueprint for writing a program to solve the problem. Once we have a blueprint of the solution, we can easily implement it in any high-level language like C, C++, Java, and so forth. It solves the problem into the finite number of steps. An algorithm written in a programming language is known as a program. A computer is a machine with no brain or intelligence. Therefore, the computer must be instructed to perform a given task in unambiguous steps. Hence, a programmer must define his problem in the form of an algorithm written in English. Thus, such an algorithm should have following features:

1. An algorithm should be simple and concise.

2. It should be efficient and effective.

3. It should be free of ambiguity; that is, the logic must be clear.

Similarly, an algorithm must have following characteristics:

- **Input:** It reads the data of the given problem.

- **Output:** The desired result must be produced.

- **Process/Definiteness:** Each step or instruction must be unambiguous.

- **Effectiveness:** Each step should be accurate and concise. The desired result should be produced within a finite time.

- **Finiteness:** The number of steps should be finite.

1.4.1 Developing an Algorithm

To develop an algorithm, some steps are suggested:

1. Defining or understanding the problem.

2. Identifying the result or output of the problem.

3. Identifying the inputs required by the problem and choosing the best input.

4. Designing the logic from the given inputs to get the desired output.

5. Testing the algorithm for different inputs.

6. Repeating the previous steps until it produces the desired result for all the inputs.

1.5   Approaches for Designing an Algorithm

A complicated algorithm is divided into smaller units which are called modules. Then these modules are further divided into sub-modules. Thus, in this way, a complex algorithm can easily be solved. The process of dividing an algorithm into modules is called modularization. There are two popular approaches for designing an algorithm:

▪ Top-Down Approach

▪ Bottom-Up Approach

Now let us understand both approaches.

1. **Top-Down Approach:** *A top-down approach states that the complex/complicated problem/algorithm should be divided into smaller modules.* These smaller modules are further divided into sub-modules. This process of decomposition is repeated until we achieve the desired output of module complexity. A top-down approach starts from the topmost module, and the modules are incremented accordingly until level is reached where we don't require any more sub-modules, that is, the desired level of complexity is achieved.

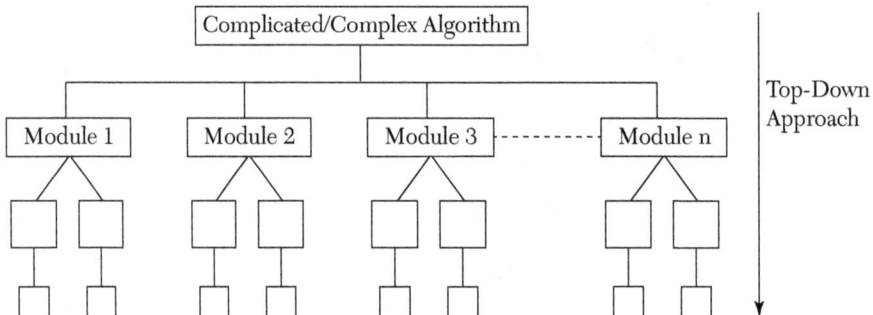

These sub modules can further be divided into one or more sub modules.

Figure 1.8. Top-down approach.

2. **Bottom-Up Approach:** A bottom-up algorithm design approach is the opposite of a top-down approach. *In this kind of approach, we first start with designing the basic modules and proceed further toward designing the high-level modules.* The sub-modules are grouped together to form a module of a higher level. Similarly, all high-level modules are grouped to form more high-level modules. Thus, this process of combining the sub-modulesis repeated until we obtain the desired output of the algorithm.

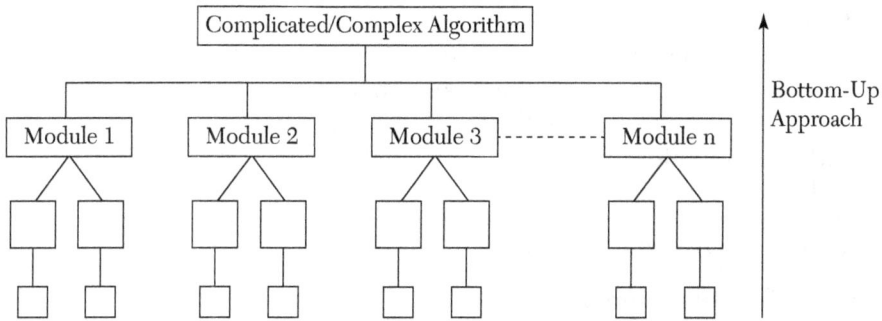

Figure 1.9. Bottom-up approach.

1.6   Analyzing an Algorithm

An algorithm can be analyzed by two factors, that is, space and time. We aim to develop an algorithm that makes the best use of both these resources. Analyzing an algorithm measures the efficiency of the algorithm. The efficiency of the algorithm is measuredin terms of speed and time complexity. The complexity of an algorithm is a function that measures the space and time used by an algorithm in terms of input size.

Time Complexity: *The time complexity of an algorithm is the amount of time taken by an algorithm to run the program completely. It is the running time of the program.* The time complexity of an algorithm depends upon the input size. The time complexity is commonly represented by using big O notation. For example, the time complexity of a linear search is $O(n)$.

Space Complexity: *The space complexity of an algorithm is the amount of memory space required to run the program completely.* The space complexity of an algorithm depends upon the input size.

Time complexity is categorized into three types:

1. **Best-Case Running Time:** The performance of the algorithm will be best under optimal conditions. For example, the best case for a binary search occurs when the desired element is the middle element of the list. Another example can be of sorting; that is, if the elements are already sorted in a list, then the algorithm will execute in best time.

2. **Average-Case Running Time:** It denotes the behavior of an algorithm when the input is randomly drawn from a given collection or distribution. It is an estimate of the running time for "average" input. It is usually assumed that all inputs of a given size are likely to occur with equal probability.

3. **Worst-Case Running Time:** The behavior of the algorithm in this case concerns the worst possible case of input instance. The worst-case running time of an algorithm is an upper bound on the running time for any input. For example, the worst case for a linear search occurs when the desired element is the last element in the list or when the element does not exist in the list.

Frequently Asked Questions

6. Define time complexity

Ans: *Time complexity is a measure which evaluates the count of the operations performed by a given algorithm as a function of the size of the input. It is the approximation of the number of steps necessary to execute an algorithm. It is commonly represented with asymptotic notation, that is, O(g) notation, also known as big O notation, where g is the function of the size of the input data.*

1.6.1 Time-Space Trade-Off

In computer science, time-space trade-off is a way of solving a particular problem either in less time and more memory space or in more time and less memory space. But if we talk in practical terms, designing such an algorithm in which we can save both space and time is a challenging task. So, we can use more than one algorithm to solve a problem. One may require less time, and the other may require less memory space to execute. Therefore, we sacrifice one thing for the other. Hence, there exists a time-space or time-memory trade-off between algorithms. Thus, this time-space trade-off gives the programmer a rational choice from an informed point of view. So, if time is a big concern for a programmer, then he or she might choose a program which takes less or the minimum time to execute. On the other hand, if space is a prime concern for a programmer, then, in that case, he or she might choose a program that takes less memory space to execute at the cost of more time.

1.7 Abstract Data Types

An abstract data type (ADT) is a popular mathematical model of the data objects which define a data type along with various functions that operate on these objects. To understand the meaning of an abstract data type, we will simply break the term into two parts, that is, "data type" and "abstract." The data type of a variable is a collection of values which a variable can take.

There are various data types in Java that include integer, float, character, long, double, and so on. When we talk about the term "abstract" in the context of data structures, it means apart from detailed specification. It can be considered as a description of the data in a structure with a list of operations to be executed on the data within the structure. Thus, an abstract data type is the specification of a data type that specifies the mathematical and logical model of the data type. For example, when we use stacks and queues, then at that point of time our prime concern is only with the data type and the operations to be performed on those structures. We are not worried about how the data will be stored in the memory. Also, we don't bother about how push () and pop () operations work. We just know that we have two functions available to us, so we have to use them for insertion and deletion operations.

1.8   Big O Notation

The performance of an algorithm, that is, time and space requirements, can be easily compared with other competitive algorithms using asymptotic notations such as big O notation, Omega notation, and Theta notation. The algorithmic complexity can be easily approximated using asymptotic notations by simply ignoring the implementation-dependent factors. For instance, we can compare various available sorting algorithms using big O notation or any other asymptotic notation.

Big O notation is one of the most popular analysis characterization schemes, since it provides an upper bound on the complexity of an algorithm. In big O, $O(g)$ is representative of the class of all functions that grow no faster than g. Therefore, if $f(n) = O(g(n))$ then $f(n) <= c(g(n))$ for all $n > n_0$ where n_0 represents a threshold and c represents a constant.

An algorithm with O(1) complexity is referred to as a constant computing time algorithm. Similarly, an algorithm with $O(n)$ complexity is referred to as a linear algorithm, $O(n^2)$ for quadratic algorithms, $O(2^n)$ for exponential time algorithms, $O(n^k)$ for polynomial time algorithms, and $O (\log n)$ for logarithmic time algorithms.

An algorithm with complexity of the order of $O(\log_2 n)$ is considered as one of the best algorithms, while an algorithm with complexity of the order of $O(2^n)$ is considered as the worst algorithm. The complexity of computations or the number of iterations required in various types of functions may be compared as follows:

$$O(\log_2 n) < O(n) < O(n \log_2 n) < O(n^2) < O(n^3) < O(2^n)$$

1.9 Summary

- A data structure determines a way of storing and organizing the data elements in the computer memory. Data means a value or a collection of values. Structure refers to a way of organizing the data. The mathematical or logical representation of data in the memory is referred as a data structure.

- Data structures are classified into various types which include linear and non-linear data structures, primitive and non-primitive data structures, static and dynamic data structures, and homogeneous and non-homogeneous data structures.

- A linear data structure is one in which the data elements are stored in a linear or sequential order; that is, data is stored in consecutive memory locations. A non-linear data structure is one in which the data is not stored in any sequential order or consecutive memory locations.

- A static data structure is a collection of data in memory which is fixed in size and cannot be changed during runtime. A dynamic data structure is a collection of data in which memory can be reallocated during execution of a program.

- Primitive data structures are fundamental data structures or predefined data structures which are supported by a programming language. Non-primitive data structures are comparatively more complicated data structures that are created using primitive data structures.

- A homogeneous data structure is one that contains all data elements of the same type. A non-homogeneous data structure contains data elements of different types.

- An array is a collection of *homogeneous* (similar) types of data elements in *contiguous* memory.

- A queue is a linear collection of data elements in which the element inserted first will be the element taken out first, that is, a FIFO data structure. A queue is a linear data structure in which the first element is inserted from one end called the REAR end and the deletion of the element takes place from the other end called the FRONT end.

- A linked list is a sequence of nodes in which each node contains one or more than one data field and an address field that stores the address of the next node.

- A stack is a linear collection of data elements in which insertion and deletion take place only at one end called the TOP of the stack. A stack is a Last In First Out (LIFO) data structure, because the last element added to the top of the stack will be the first element to be deleted from the top of the stack.

- A tree is a non-linear data structure in which the data elements or the nodes are represented in a hierarchical order. Here, an initial node is designated as the root node of the tree, and the remaining nodes are partitioned into two disjointed sets such that each set is a part of a subtree.

- A binary tree is the simplest form of a tree. A binary tree consists of a root node and two subtrees known as the left subtree and right subtree, where both the subtrees are also binary trees.

- A graph is a general tree with no parent-child relationship. It is a non-linear data structure which consists of vertices or nodes and the edges which connect those vertices with one another.

- An algorithm is a systematic set of instructions combined to solve a complex problem. It is a step-by-finite-step sequence of instructions, each of which has a clear meaning and can be executed in a minimum amount of effort in finite time.

- The process of dividing an algorithm into modules is called modularization.

- The time complexity of an algorithm is described as the amount of time taken by an algorithm to run the program completely. It is the running time of the program.

- The space complexity of an algorithm is the amount of memory space required to run the program completely.

- An ADT (Abstract Data Type) is a mathematical model of the data objects which define a data type as well as the functions to operate on these objects.

- Big O notation is one of the most popular analysis characterization schemes, since it provides an upper bound on the complexity of an algorithm.

1.10   Exercises

1.10.1 Theory Questions

1. How do you define a good program?

2. Explain the classification of data structures.

3. What is an algorithm? Discuss the characteristics of an algorithm.

4. What are the various operations that can be performed on data structures? Explain each of them with an example.

5. Differentiate an array with a linked list.

6. Explain the terms time complexity and space complexity.

7. Write a short note on graphs.

8. What is the process of modularization?

9. Differentiate between stacks and queues with examples.

10. What is meant by abstract data types (ADT)? Explain in detail.

11. How do you define the complexity of an algorithm? Discuss the worst-case, best-case, and average-case time complexity of an algorithm.

12. Write a brief note on trees.

13. Explain how you can develop an algorithm to solve a complex problem.

14. Explain time-memory trade-off in detail.

1.10.2 Multiple Choice Questions

1. Which of the following data structures is a FIFO data structure?
   (a) Array
   (b) Stacks
   (c) Queues
   (d) Linked List

2. How many maximum children can a binary tree have?
   (a) 0
   (b) 2
   (c) 1
   (d) 3

3. Which of the following data structures uses dynamic memory allocation?
   (a) Graphs
   (b) Linked Lists
   (c) Trees
   (d) All of these

4. In a queue, deletion is always done from the _____.

   (a) Front end

   (b) Rear end

   (c) Middle

   (d) None of these

5. Which data structure is used to represent complex relationships between the nodes?

   (a) Linked Lists

   (b) Trees

   (c) Stacks

   (d) Graphs

6. Which of the following is an example of a heterogeneous data structure?

   (a) Array

   (b) Structure

   (c) Linked list

   (d) None of these

7. In a stack, insertion and deletion takes place from the _____.

   (a) Bottom

   (b) Middle

   (c) Top

   (d) All of these

8. Which of the following is not part of the Abstract Data Type (ADT) description?

   (a) Operations

   (b) Data

   (c) Both (a) and (b)

   (d) None of the above

9. Which of the following data structures allows deletion at one end only?

   (a) Stack

   (b) Queue

   (c) Both (a) and (b)

   (d) None of the above

10. Which of the following data structures is a linear type?
   (a) Trees
   (b) Graphs
   (c) Queues
   (d) None of the above

11. Which one of the following is beneficial when the data is stored and has to be retrieved in reverse order?
   (a) Stack
   (b) Linked List
   (c) Queue
   (d) All of the above

12. A binary search tree whose left and right subtree differ in height by 1 at most is a _____.
   (a) Red Black Tree
   (b) M way search tree
   (c) AVL Tree
   (d) None of the above

13. The operation of processing each element in the list is called _____.
   (a) Traversal
   (b) Merging
   (c) Inserting
   (d) Sorting

14. Which of the following are the two primary measures of the efficiency of an algorithm?
   (a) Data & Time
   (b) Data & Space
   (c) Time & Space
   (d) Time & Complexity

15. Which one of the following cases does not exist/occur in complexity theory?
   (a) Average Case
   (b) Worst Case
   (c) Best Case
   (d) Minimal Case

CHAPTER 2

INTRODUCTION TO THE JAVA LANGUAGE

2.1 Introduction

Java is a widely used programming language that was originally created by James Gosling, Mike Sheridan, Chris Warth, Ed Frank, and Patrick Naughton at Sun Microsystems in 1991. Sun Microsystems released the first public implementation of Java 1.0 in 1996. The primary motivation for developing Java was the need for a platform-independent language that could be used to create software that could be embedded in different electronic devices, such as washing machines, remote controls, televisions, and so on. There are various types of CPUs available which are used as controllers that are mainly designed to be compiled only for a specific target with the help of C, C++, Java, and other programming languages. But one limitation is the time-consuming process of creating the compilers; in addition, the compilers are too expensive. Hence, to overcome this limitation and to find an easier and more cost-efficient solution, James Goslings along with his coworkers began their work on a portable and platform-independent language that could be used to produce code that would run on a variety of CPUs under different environments, which ultimately led to the creation of the Java language. Mainly, there were three primary goals in the creation of the Java language, which are as follows:

- It must be simple, object-oriented, and a familiar language.

- It must be portable, distributed, and platform independent.

- It must be robust, secure, and reliable.

2.2 Java and Its Characteristics

Although Java derives many of its characteristics from C and C++, it is not an enhanced version of C++. It uses the syntax of the C language and echoes the object-oriented features of the C++ language. Java is a very powerful language and it is known as a language for professional programmers. There are several different characteristics of the Java programming language. Some of them are as follows:

- Java is a very simple language that is very easy to learn, and its syntax is simple and easy to understand.

- Java is an object-oriented programming language. Everything in Java is an object.

- Java is best known for its *security*. Since everything in Java is considered as an object, it increases its security. Also, Java is secure because it does not use explicit pointers and because Java programs run inside a virtual machine sandbox.

- Java is a *robust* language because it uses strong memory management concepts and contains an automatic garbage collection mechanism. It also supports powerful exception handling. All these features make Java robust.

- Java is *portable* because the Java code can be run on any platform. It doesn't require any further implementation.

- Java is *faster* than other traditional interpreted programming languages because Java bytecode is very "close" to the native code, and thus it offers high performance.

- Java is a *distributed*, *multi-threaded*, and *dynamic* language.

2.3 Java Overview

In Java, initially all the source code is written into plaintext files ending with the .java (dot java) extension. All the source files are then compiled into .class files by the java c compiler. A .class file does not contain the code that is native to your processor; instead it contains bytecode, which is the machine language of the Java Virtual Machine (JVM). Finally, the java launcher tool then runs the application with an instance of the JVM. As the JVM is available on different operating systems, the same .class files are capable of running on different operating systems like Windows, macOS, Linux, and so forth.

Source code

Source code files are converted into bytecode files

Bytecode

Through JVM, same application is capable of running on different platforms

Windows

JAVAC Compiler

JAVAC VM

Hello.java file

Hello.class file

Hello

Unix

MacOs

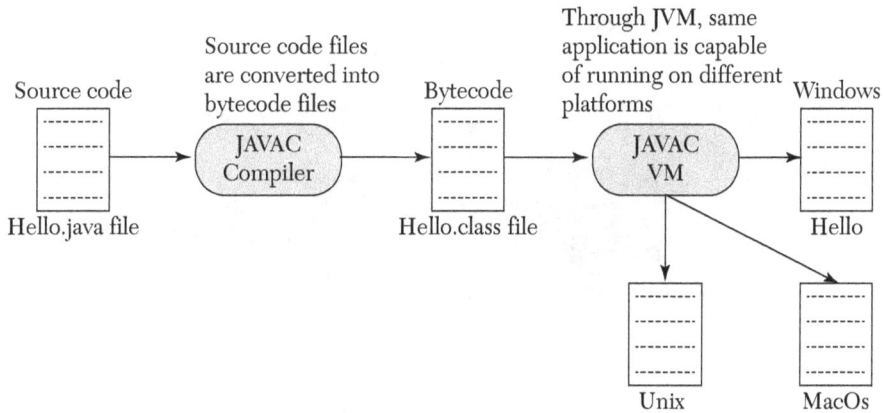

Figure 2.1. Overview of Java SE process.

2.4   Compiling the Java Program

Each file in Java uses the .java filename extension. Also, in Java all the code must be inside a class. The most important task is that, by convention, the name of the class should match the name of the file that holds the program.

To compile a program, open the command prompt in your system. Run the compiler javac (Java compiler), specifying the source file on the following command prompt:

```
Administrator: Command Prompt                    —   □   ✕

C:\Program Files\Java\jdk-12.0.2\bin>javac Filename.java
```

The source file is compiled into a .class file by the compiler. Now, the Filename.class file will be created, which will contain the bytecode of the program. To run the program, use the Java Launcher tool also known as java.

```
Administrator: Command Prompt                    —   □   ✕

C:\Program Files\Java\jdk-12.0.2\bin>javac Filename.java

C:\Program Files\Java\jdk-12.0.2\bin>java Filename
```

Finally, when the program is executed, the output is displayed as:

```
Administrator: Command Prompt                        —    □    ×

C:\Program Files\Java\jdk-12.0.2\bin>javac Filename.java

C:\Program Files\Java\jdk-12.0.2\bin>java Filename
Hello World!!!

C:\Program Files\Java\jdk-12.0.2\bin>
```

2.5   Object-Oriented Programming

Object-oriented programming is the core of the Java programming language. Object-oriented programming (OOP) is so integral to Java that it is best to understand its basic principles. OOP organizes a program around its data and a set of well-defined interfaces to that data. The following are the various features or characteristics of object-oriented programming:

- **Objects:** An *object* is a container in which the data is stored in a combination of variables, methods, and data structures. An object is that component of a program that interacts with the other parts/pieces of the program. Objects are the fundamental runtime entities of a program.

- **Classes:** The building block of Java that leads to object-oriented programming is a *class*. A class is actually a user-defined data type that holds its own data members and member methods. It is a way to bind the data members and methods together. A class can be accessed by creating an object of that class. Classes provide a convenient method for packing together a group of logically related data items and methods that work on them.

- **Encapsulation:** *Encapsulation* is a process of wrapping up data members and member methods into a single unit. It is the mechanism which keeps the code and the data safe from external interference. It implies that there is no direct access granted to the data; that is, it is hidden. So, in order to access that data, we must interact with the object that is responsible for the data. This is also known as *data hiding*. The process of encapsulation makes the data of the system more secure and reliable.

Practical Application

The most common example of encapsulation is a capsule. In a capsule, all the medicines are encapsulated inside a single capsule.

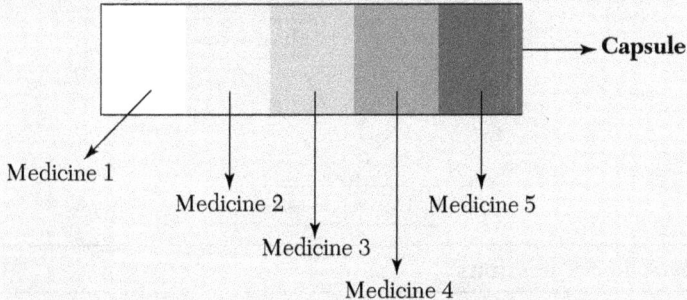

- **Inheritance:** Inheritance is the process of deriving a new class from the existing one. The existing class is known as the "base class," "parent class," or "superclass." The new classis known as the "derived class," "child class," or "subclass." The process of inheritance allows the child classes to inherit all the features, that is, variables and methods, of their parent class. Thus, the derived classes will have all the features of their base class, and the programmer can add some new features specific to the derived class. There are three types of inheritances supported in Java, which are as follows:

1. **Single Inheritance:** A class derived from a single base class is known as a single inheritance.

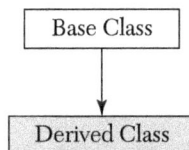

2. **Multilevel Inheritance:** Classes derived from the already derived classes are known as multilevel inheritances.

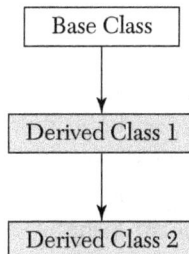

3. **Multiple Inheritance:** A class derived from more than one base class is known as a multiple inheritance. It is not achieved directly through classes in Java. It is achieved with the help of interfaces which will be covered in the upcoming chapters.

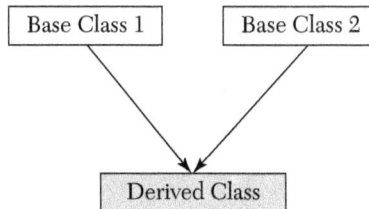

```
┌──────────────┐        ┌──────────────┐
│ Base Class 1 │        │ Base Class 2 │
└──────────────┘        └──────────────┘
          \                  /
           \                /
            ┌──────────────────┐
            │  Derived Class   │
            └──────────────────┘
```

Frequently Asked Questions

1. What is the difference between the base and derived class?

Ans: *When creating a class, instead of writing completely new data members and member methods, the programmer can designate that the new class should inherit the members of an existing class. This existing class is known as a base class, and the new class is known as a derived class. A class can be derived from more than one class, which means it can inherit data and methods from multiple base classes.*

Practical Application

The most common real-life example of inheritance is your family, in which your grandfather is one head of the family (base class/parent class), your father is the child of the grandfather (derived class/child class), and you are the child of your father (another derived class).

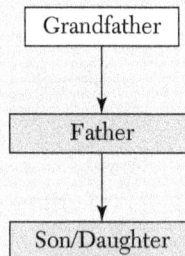

```
        ┌──────────────┐
        │ Grandfather  │
        └──────────────┘
               │
               ▼
        ┌──────────────┐
        │   Father     │
        └──────────────┘
               │
               ▼
        ┌──────────────┐
        │ Son/Daughter │
        └──────────────┘
```

- **Polymorphism:** The word *polymorphism* is derived from the word *poly*, which means many, and the word *morph*, which means forms. Therefore, anything that exists in more than one form is referred to as a polymorph. Thus, polymorphism means the ability to make more than one form. Polymorphism usually occurs when there is a hierarchy of classes and the classes are related by inheritance. Polymorphism is considered as one of the most important features of object-oriented programming. Polymorphism is of two types:

1. **Compile Time Polymorphism (Static Polymorphism):** It is done using method overloading.

2. **Runtime Polymorphism (Dynamic Polymorphism):** It is done using method overriding.

Practical Application

A person exhibits different roles in different situations; that is, he/she is a child for his/her father, he/she is a student for his/her teachers, he/she is a friend for his/her friends, and so on.

- **Abstraction:** The process of *abstraction* is somewhat related to the idea of hiding data that is not essential. Abstraction refers to the act of displaying only the essential features and hiding the background details and explanations that are not needed for presentation. This is also known as data abstraction. Abstraction is one of the vital features provided by the Java programming language. In Java, abstraction is achieved using abstract classes and interfaces.

Practical Application

Consider a man driving a bus. The man only knows that pressing the accelerator will increase the speed of the bus and that applying the brakes decreases the speed of the bus. The man does not know how pressing the accelerator increases the speed or how applying the brakes decreases the speed. Hence, he does not know the inner mechanism of the bus.

Frequently Asked Questions

2. Explain the differences between abstraction and encapsulation.

Ans: *Data abstraction refers to providing only essential information to the outside world and hiding the background details, that is, representing the needed information in the program without presenting all the details. On the other hand, encapsulation is an object-oriented programming concept that binds together the data and the methods that manipulate the data and thus keeps them safe from the outside world.*

- **Message Passing:** Message passing is the process of communication between objects. An object-oriented program consists of objects that communicate by sending and receiving information from each other. A message for an object is a request for the execution of a procedure, and thus it will invoke a method in the receiving object that generates the desired results. Message passing involves three things: the name of the object, the name of the method, and the data/information to be sent.

■ **Dynamic Binding:** Dynamic binding is the process by which the code that has to be executed for a given procedure call is known at runtime rather than at the compile time. It is also known as *dynamic dispatch*. It is an object-oriented programming concept, and it is also related to inheritance and polymorphism.

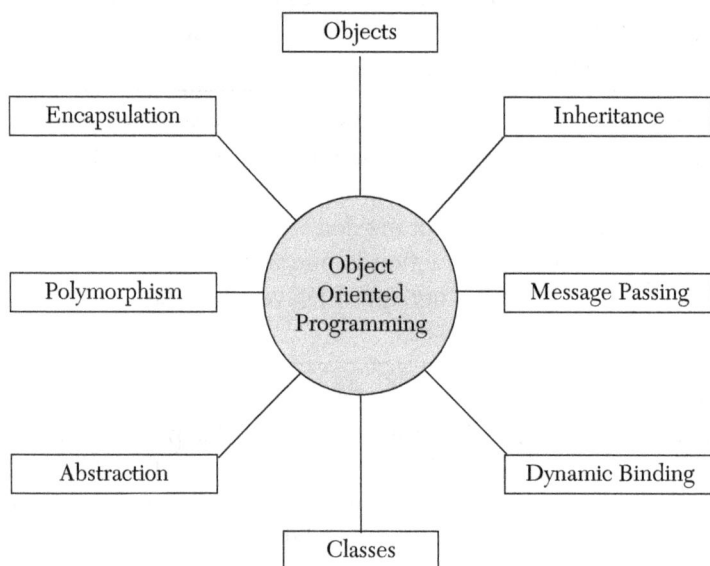

Figure 2.2. Features of object-oriented programming.

2.6   Character Set Used in Java

The character set allowed in Java consists of the following characters:

■ **Alphabet:** It includes uppercase as well as lowercase letters of English, i.e., {A, B, C... ., Z} and {a, b, c... ., z}.

■ **Digits:** It includes decimal digits, i.e., {0, 1, 2 . . ., 9}.

■ **White Spaces:** It includes spaces, enters, and tabs.

■ **Special Characters:** It consists of special symbols which include {, !, ?, #, <, >, (,), %, ", &, ^, *, <<, >>, [,], +, =, /, -, _, :, ;, }.

2.7 Java Tokens

Java tokens help us to write a program in Java. Java supports various types of tokens:

- Keywords
- Identifiers
- Constants
- Variables

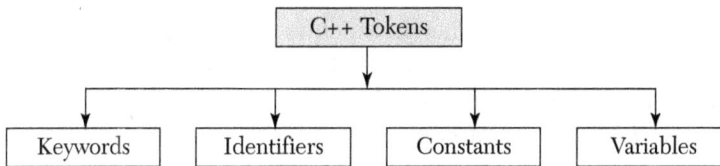

Figure 2.3. Various Java tokens.

Now, let us discuss all of them.

Keywords: *Keywords in Java are the reserved words which have a special meaning.* They are written in lower case. *Keywords cannot be used as identifiers.* Examples are `auto`, `int`, `float`, `char`, `break`, `continue`, `double`, `long`, `short`, `while`, `for`, `else`, `new`, `void`, and so forth.

Identifiers: *An identifier is a name which is given to a constant, variable, method, or array.* The rules which are used to define identifiers are as follows:

1. An identifier can have letters, digits, or underscores.

2. It should not start with a digit.

3. It can start with an underscore or a letter.

4. An identifier cannot have special symbols or spaces.

5. A keyword cannot be used as an identifier.

For example:

Acceptable Identifiers	Unacceptable Identifiers
a345_	au to
yourname34	12d
c_65	n 3_
average	5rf_t3

Constants: *Constants are the fixed values in Java that can never be changed.* Constants are also known as literals. For example, the value of pi is always fixed. A constant can be of any basic data type. Java has two types of constants, which are as follows:

- **Numeric Constant:** It is a constant to which only the integer values are assigned. It consists of an integer constant, float constant, or real constant.

- **Character Constant:** It is a constant to which only the character values are assigned.

Variables: A variable is a name which is used to refer to some memory location. While working with a variable, we refer to the address of the memory where the data is stored. Java supports two types of variables, which include character variables and numeric variables.

- **Numeric Variables:** These are used to store the integer or floating-type values.

- **Character Variables:** In this variable, single characters are enclosed in single quotes.

2.8 Data Types in Java

Data types are the special keywords which define the type of data and the amount of data a variable is holding. There are five primitive data types available in Java, which are as follows:

- **Integer (int)** that is, 23, -98, +786, etc.

- **Character (char)**, that is, 'A', 'x', 'u', etc.

- **Boolean**, that is, either true/false or 1/0.

- **Floating point (float)** that is, 1.7, -4.6, +9.6, etc.

- **Double floating point (double)**, that is, -87.55653, +9867.3467, etc.

There are various modifiers which are used to alter the meaning of the basic data types, except for void. These modifiers are signed, unsigned, short, and long. All these modifiers can be applied to integers. However, only signed and unsigned can be applied to characters. Further classification is shown in the following table.

Data Type	Bytes (in memory)	Range
int	2	-2,147,483,648 to 2,147,483,648
char	1	-128 to 127
double	8	4.9e-324 to 1.8e+308
float	4	1.4e-045 to 3.4e+038
short	2	-32,768 to 32,767
long	4	-9,223,372,036,854,775,808 to 9,223,372,036,854,775,808
boolean	1	true or false

2.9 Structure of a Java Program

Unlike C and C++, a Java program needs a class to be executed. In C and C++, the executable code or the structure may or may not have a class. But in Java everything is confined within a class.

Now, let us see it with the help of a simple program.

```
class Example     /*A class is created having data
                     members and member methods*/
{
 private int a, b, sum; /*a, b & sum are private data
                     /members of the class*/
 public void sum()/*sum() method is declared as public*/
 {
   a=10,b=20;
   sum = a + b;

   System.out.println("Sum" + sum); /*Prints the output on screen*/
 }
public static void main(String[] args) /*Main Method*/
{
   Example obj = new Example(); /*obj is the object of the class
                     /Example*/
   obj.sum();
}
```

Frequently Asked Questions

3. How is input taken and how is output displayed in Java? Explain.

Ans: *In Java, the Scanner class is used to take input. It is a predefined class and can be accessed by creating an object of it. The output is displayed using the statement 'System.out.println();'.*

```
Example:
import java.util.Scanner; /*importing Scanner Class*/
public class ABC
{
   public static void main(String[]args)
{

   Scanner obj=new Scanner(System.in);
   System.out.println("Enter a value:");
   int x =scn.nextInt();
   System.out.println(x);
   }
}
```

2.10   Operators in Java

Operators in Java are used to perform some specific operations between the different variables and constants. Java supports a variety of operators, which are given as follows:

- Arithmetic Operators
- Logical Operators
- Assignment Operators
- Relational Operators/ Comparison Operators
- Condition Operators/ Ternary Operators
- Bitwise Operators
- Unary Operators

Now, let us discuss all of these operators.

Arithmetic Operators

Arithmetic operators are those operators which are used in mathematical computation or calculation. The valid arithmetic operators in Java are given in the following table.

Let x and y be the two variables.

Operator	Operation	Example
+	Addition	x + y
−	Subtraction	x − y
*	Multiplication	x * y
%	Remainder/ Modulus	x % y
/	Division	x / y

Logical Operators

Java supports three types of logical operators, which are given as follows:

Operator	Description	Example
!	Logical NOT	!x, !y
&&	Logical AND	x && y
||	Logical OR	x || y

Logical NOT: It is a unary operator. This operator takes a single expression, and it inverts the result such that true becomes false and vice versa. The truth table for logical NOT is given as follows:

X	Y	!x	!y
0	1	1	0
1	0	0	1

Logical AND: It is a binary operator. Hence, it takes two inputs or expressions. If both the inputs are true, then the whole expression is true. If both or even any one of the inputs is false, then the whole expression will be false. The truth table for logical AND is given as follows:

X	Y	X && Y
0	0	0
0	1	0
1	0	0
1	1	1

Logical OR: It is also a binary operator; that is, it also takes two expressions. If both the inputs are false, then the output is false. If a single input or both of the inputs are true, then the output will be true. The truth table for logical OR is given as follows:

X	Y	X ‖ Y
0	0	0
0	1	1
1	0	1
1	1	1

Assignment Operators

Assignment operators are ones which are responsible for assigning values to the variables. These operators are always evaluated from right to left. Java supports various assignment operators, which are given in the following table:

Operators	Example
=	x = 5, y = 8
+=	x += y :− x = x + y
−=	x −= y :− x = x − y
*=	x *= y :− x = x * y
%=	x %= y :− x = x % y
/=	x /= y :− x = x / y

Relational Operators

Relational operators are used for comparison between two values or expressions. They are also known as comparison operators. These operators are always evaluated from left to right. The various relational operators used in Java are as follows:

Operators	Description	Example
>	Greater than	x > y
<	Less than	x < y
==	Equal to	x == y
>=	Greater than equal to	x >= y
<=	Less than equal to	x <= y
!=	Not equal to	x != y

Conditional Operators

The conditional operator is also known as a ternary operator regarding input; it accepts three operands. The syntax of this operator is as follows:

Syntax – (Expression 1)? (Expression 2) :(Expression 3);

Where, expression 1 is first evaluated. If expression 1 is true, then expression 2 is evaluated and expression 2 will be the answer of this whole expression, else expression 3 is evaluated and expression 3 will be the answer of this whole expression. Conditional operators can be used to find the larger of two numbers.

Greatest = (x > y)? X: y;

Here, if x > y is true, then x is greater than y; that is, (greatest = x) else y is greater than x, that is, (greatest = y).

Bitwise Operators

Bitwise operators are the special operators that are used to perform operations at the bit level. Java supports various types of bitwise operators, which include the following:

Operator	Description
<<	Right shift
>>	Left shift
&	Bitwise AND
		Bitwise OR
^	Bitwise XOR

Unary Operators

A unary operator is one which requires only a single operand to work. Java supports two unary operators, which are increment (++) and decrement (--) operators. These operators are used to increase or decrease the value of a variable by one respectively. The two variants of increment and decrement operators are postfix and prefix. In a postfix expression, the operator is applied after the operand. On the other hand, in a prefix expression, the operator is applied before the operand.

Operator	Postfix	Prefix
Increment (++)	x++	++x
Decrement (--)	--x	--x

Remember, x++ is not the same as ++x; in x++ the value is returned first, and then the value is incremented. In ++x, the value is returned after it is incremented. Similarly, x-- is not the same as --x. It is true that both these operators increment or decrement the value by 1. For example,

b = a++, is equivalent to:

1. b = a,

2. a = a + 1.

Similarly, b = --a is equivalent to:

1. a = a – 1,

2. b = a.

2.11 Decision Control Statements in Java

Whenever we talk of a program written in the Java language, we know that a Java program will always execute sequentially, that is, line by line. Initially, the first line will be executed. Then, the second line will execute after the execution of the first line, and so on. *Control statements* are those that enable a programmer to execute a particular block of code and specify the order in which the various instructions of code are required to be executed. They determine the flow of control. Control statements define how the control is transferred to other parts of a program. Hence, they are also called decision control statements. A decision control statement is one that helps us to jump from one point of a program to another. A decision control statement is executed in Java using the following:

- IF statement

- IF-ELSE statement

- Nested IF-ELSE statement

- Switch case statement

Now, let us discuss all of them.

IF Statement

An IF statement is a bidirectional control statement which is used to test the condition and take one of the possible actions. It is the simplest decision control statement and is used very frequently in decision making. The general syntax of an **IF** statement is as follows:

```
if (condition)
{
 Statement Block of if;    //If condition is true, execute the
statements of if.
}
Statements Block under if;
```

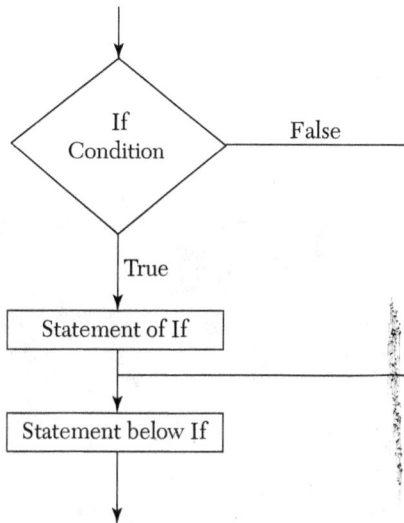

Figure 2.4. IF statement flow diagram.

The IF statement will check the condition, and if the condition is true, then only the set of statements inside the IF block will be executed. However, the set of statements outside the IF block will always be executed independently of the condition. An IF block can have one or multiple statements enclosed within curly brackets. Also, the ELSE block is optional in a simple IF statement because if the condition is false, then the control directly jumps to the next statement. Remember, there is no semicolon after the condition, because the condition and the statement should be used as a single statement.

For example:

//Write a program to show the use of the IF statement.

```
import java.util.*;
public class If
{
```

```
    public static void main(String[]args)
{

    Scanner src=new Scanner(System.in);
    System.out.println("Enter two values:");
    int x=src.nextInt();
    int y=src.nextInt();
    if(y>x)
{

        System.out.println("Y is greater:"+y);
    }
        System.out.println("End of the program");
    }
}
```

The output of the program is shown as:

IF-ELSE Statement

After discussing the usage of the IF statement, we learned that the IF statement does nothing when the condition is false. It just passes the control to the next statement outside of the IF block. The IF-ELSE statement takes care of this aspect. The general syntax of the IF-ELSE statement is as follows:

```
if (condition)
{
    Statements X;  //If condition is true, execute the statements X
}
else
{
    Statements Y;  //If condition is false, execute the statements Y
}
```

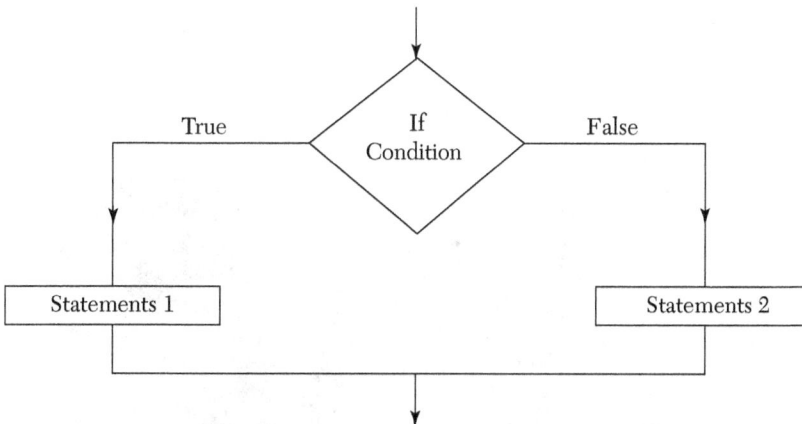

Figure 2.5. IF-ELSE statement flow diagram.

The IF-ELSE statement will check the condition. If the condition is true, then the set of statements X is executed, and the ELSE block is not executed. Otherwise, if the condition is false, then the set of statements Y is executed, and the IF block is not executed. The IF or ELSE blocks can contain one or multiple statements.

For example:

//Write a program to show the use of IF-ELSE.

```java
import java.util.*;
public class IfElse
{
    public static void main(String[] args)
    {
    Scanner src = new Scanner(System.in);
    System.out.println("Enter two values:");
    int x = src.nextInt();
    int y = src.nextInt();
    if(y>x)
    {
        System.out.println("Y is greater:" +y);
    }
    else
    {
        System.out.println("X is greater:" +x);
    }

    System.out.println("End of the program");
```

```
    }

}
```

The output of the program is shown as:

Nested IF-ELSE Statement

The nested IF-ELSE statement is also known as the IF-ELSE-IF Ladder. The IF-ELSE-IF statement works the same way as that of a normal IF statement. The general syntax of the nested IF-ELSE statement is as follows:

```
if (condition 1)
{
    Statements 1;    //If condition 1 is true, execute the
                     //statements of If.
}
else if (condition 2)
{
    Statements 2;    //If condition 2 is true, execute the
                     //statements of else if.
}
else
{
    Statements 3;    //If condition 2 is false, execute the
                     //statements of else.
}
```

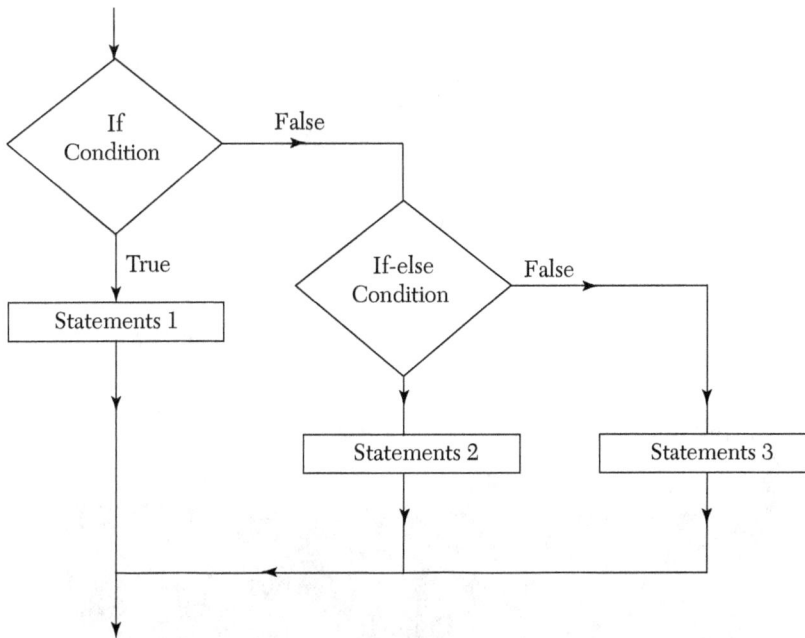

Figure 2.6. Nested IF-ELSE statement flow diagram.

An IF-ELSE-IF ladder works in the following manner. First, the IF condition is checked. If the condition is true, then the set of statements 1 is executed. If the condition is false, then the ELSE-IF condition is checked. If the ELSE-IF condition is true, then the set of statements 2 is executed. Otherwise, the set of statements 3 is executed. Remember, after the first IF expression, we can have as many ELSE-IF branches as are needed, depending upon the number of expressions to be tested.

For example:

// Write a program to show the use of Nested IF-ELSE.

```
import java.util.*;
public class IfElseIf
{
    public static void main(String[] args)
    {
        Scanner src = new Scanner(System.in);
        System.out.println("Enter three values:");
        int x = src.nextInt();
        int y = src.nextInt();
        int z = src.nextInt();
        if(x > y && x > z)
        {
```

```
        System.out.println("X is greatest:" +x);
    }
    else if(y > x && y > z)
    {
        System.out.println("Y is greatest:" +y);
    }
    else
    {
        System.out.println("Z is greatest:" +z);
    }
        System.out.println("End of the program");
    }
}
```

The output of the program is shown as:

SWITCH Statement

As we all know, an IF statement is used to check the given condition and choose one option, depending upon whether the condition is true or false. But if we have several options to choose from, then it will not be a good thing to use if statements for each option, as it will become very complex. Hence, to avoid such a problem, a switch statement is used. *The switch statement is a multidirectional conditional control statement. It is a simplified version of an IF-ELSE-IF statement. It selects one option from the number of options available to us. Thus, it is also known as a selector statement.* Its execution is faster than an IF-ELSE-IF construct. Also, a switch statement is comparatively easy to understand and debug. The general syntax of the switch statement is as follows:

```
switch (choice)
{
```

```
  case constant 1:
  Statements 1 ;
break ;
  case constant 2:
  Statements 2 ;
break ;
  case constant 3:
  Statements 3 ;
break ;
   .
   .
   .
  case constant n:
  Statements n ;
break ;
  default:
  Statements D ;
}
```

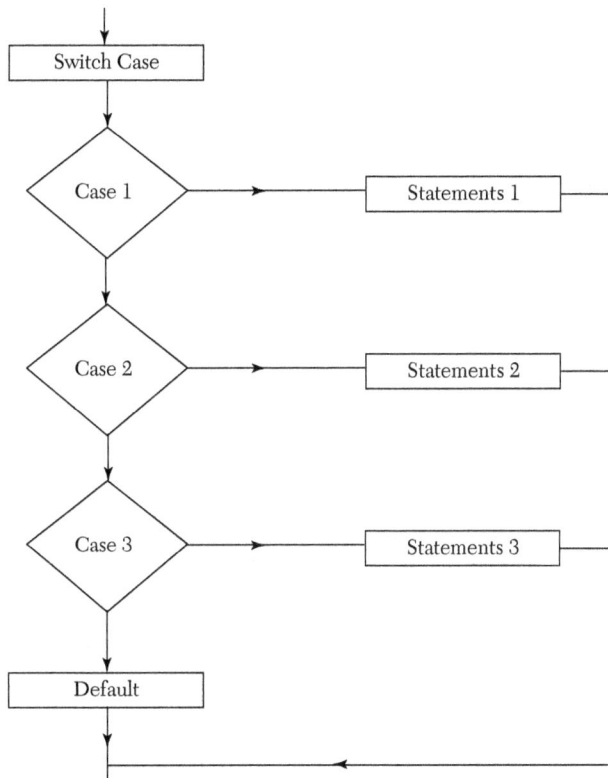

Figure 2.7. Switch statement flow diagram.

A switch statement works as follows:

- Initially, the value of the expression is compared with the case constants of the switch construct.

- If the value of the expression and the switch statement match, then its corresponding block is executed until a break is encountered. Once a break is encountered, the control comes out of the switch statement.

- If there is no match in the switch statements, then the set of statements of the default is executed.

- All the values of the case constants must be unique.

- There can be only one default statement in the entire switch statement. A default statement is optional; if it is not present and there is no match with any of the case constants, then no action takes place. The control simply jumps out of the switch statement.

For example:

```java
//Write a program to demonstrate the switch case.
import java.util.*;
public class Switch
{
    public static void main(String[] args)
    {
        Scanner src = new Scanner(System.in);
        System.out.println("Enter your choice:");
        int choice = src.nextInt();
        switch(choice)
        {
            case 1:
                System.out.println("Inside First Case!");
                break;
            case 2:
                System.out.println("Inside Second Case!");
                break;
            case 3:
                System.out.println("Inside Third Case!");

                break;
            default:
                System.out.println("Wrong choice");
```

```
          }

     }

}
```

The output of the program is shown as:

Frequently Asked Questions
4. Which one is better—a switch case or an else-if ladder?

Ans:

1. *The switch permits the execution of more than one alternative, whereas an IF statement does not. Various alternatives in an IF statement are mutually exclusive, whereas alternatives may or may not be mutually exclusive within a switch statement.*

2. *A switch can only perform equality tests involving integer type or character type constants; an if statement, on the other hand, allows for a more general comparison involving other data types as well.*

When there are more than three or four conditions, use the switch case rather than a long nested if statement.

2.12   Looping Statements in Java

Looping statements, also known as iterative statements, are the sets of instructions which are repeatedly executed until a certain condition or expression becomes false. This kind of repetitive execution of the statements in a program is called a loop. Loops can be categorized into two categories: *pre-deterministic* loops and *deterministic* loops. Pre-deterministic loops are ones in which the number of times a loop will execute is known. On the contrary, loops in which the number of times they will execute is not known are called deterministic loops. Java supports three types of loops, which include the following:

- While Loop
- Do-while Loop
- For Loop

Now, let us discuss these loops in detail.

WHILE Loop

A while loop is a loop which is used to repeat a set of one or more instructions/ statements until a particular condition becomes false. In a while loop, the condition is checked before executing the body of the loop or any statements in the statements block. Hence, a while loop is also called an entry control loop. A while loop is a deterministic loop, as the number of times it will execute is known to us. The general syntax of a while loop is as follows:

```
while (condition)
{
  block of statements/ body of loop ;
  increment/ decrement ;
}
```

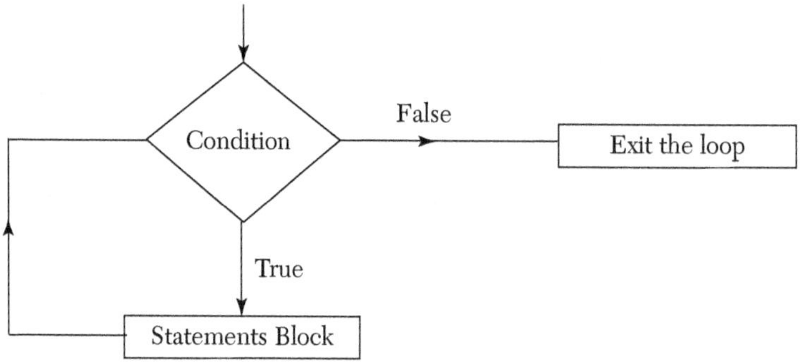

Figure 2.8. While loop flow diagram.

A while loop is executed as follows:

1. The condition is tested.

2. If the condition is true, then the statement is executed and step 1 is repeated.

3. If the condition is false, then the loop is terminated, and the control jumps out to execute the rest of the program.

For example:

//Write a program to execute a while loop.

```
public class While
{
    public static void main(String[] args)
    {
        inti = 1;
        while(i< 10)
        {
                System.out.println(i);
                i = i + 1;
        }
    }
}
```

The output of the program is shown as:

In the previous example, i is initialized to 1 and 1 is less than 10, and therefore the condition is true. Hence, the value of i is printed and is incremented by 1. The condition will become false when i becomes 10, thus at that condition, the loop will end.

DO-WHILE Loop

A do-while loop is similar to a while loop. The only difference is that, unlike a while loop in which a condition is checked at the start of the loop, in a do-while loop the condition is checked at the end of the loop. Hence, it is also called an exit control loop. This implies that in a do-while loop the statements must be executed at least once, even if the condition is false, because the condition is checked at the end of the loop. The general syntax for a do-while loop is as follows:

```
Do
{
   block of statements/ body of loop ;
   increment/ decrement ;
 } while (condition);
```

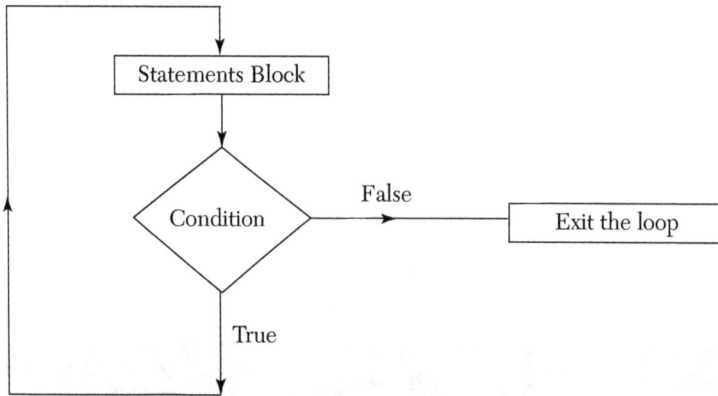

Figure 2.9. Do-while loop flow diagram.

The do-while loop continues to execute until the condition evaluates to false. A do-while loop is usually employed in situations where we require the program to be executed at least once, such as in menu-driven programs. One of the major disadvantages of using a do-while loop is that a do-while loop will always execute at least once even if the condition is false. Therefore, if the user enters some irrelevant data, it will still execute.

For example:

// Write a program to illustrate a do-while loop.

```
public class dowhile
{
    public static void main(String[] args)
    {
        int i = 0;
        do
        {
            System.out.println(i);
            i = i + 1;
```

```
        }while(i< 10);
    }
}
```

The output of the program is shown as:

In the previous code, i is initialized to 0, so the value of i is printed and is incremented by 1. After executing the loop once, now the condition will be checked. Now i = 1 and the condition is true. Therefore, the loop will execute. The condition will become false when i becomes equal to 10. In that case, the loop will be terminated.

FOR Loop

A for loop is a pre-deterministic loop; that is, it is a count-controlled loop such that the programmer knows in advance how many times the for loop is to be executed. In a for loop, the loop variable is always initialized exactly once. The general syntax of a for loop is as follows:

```
for (initialization; condition; increment/decrement)
{
  block of statements/ body of loop ;
}
```

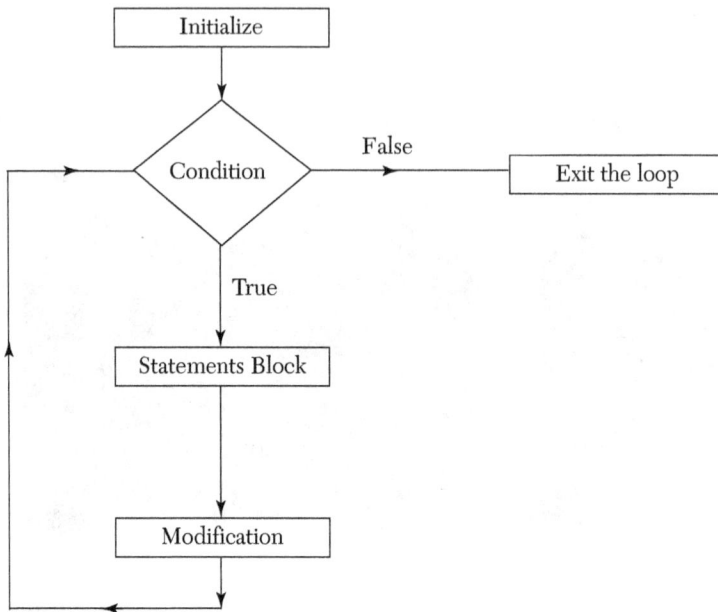

Figure 2.10. For loop flow diagram.

In a for loop, the condition is always checked at the top of the loop. Also, with every iteration of the loop, the variable and the condition are checked. If the condition is true, then the statements written within the for loop are executed; otherwise, the control moves out of the loop, and the for loop is terminated. As we have seen in the syntax, initialization means to assign a particular value to a variable initially. Second, the condition specifies whether the loop will continue to execute or will terminate. The condition is checked with every iteration of the loop. Iteration means to update the value of a variable either by incrementing it or decrementing it. Also, each section in a for loop is separated by a semicolon. So, it is possible that one of the sections may be empty. For loops are widely used to execute a particular set of statements a limited number of times.

For example:

```
// Write a program to demonstrate a for loop.
public class For
{
    public static void main(String[] args)
    {
        for(int i = 1 ; i<= 10 ; i++)
        {
```

```
            System.out.println(i);
        }
    }
}
```

The output of the program is shown as:

In the previous example, i is a counter variable which is initialized to 1. Now, the condition is checked because 1 is less than 10. Thus, the condition is true, so the value of i is printed. After every iteration, the value of i is incremented, and the condition is checked. The condition will become false when i becomes 11, so at that time the for loop will be terminated, and the control will come out of the loop.

2.13 Break and Continue Statements

In Java, *break* statements are used for loops and switch statements. They are used to terminate the execution of the loop. *A break statement causes an intermediate exit from the loop in which the statement appears.* We have already seen its use in switch statements, as it is used to exit from a switch statement. When a break is encountered, the control jumps out of the loop. The break statement is usually used in a situation in which either there is some error or if we don't want to execute the rest of the loop. It has a very simple syntax:

Syntax – `break;`

For example:

//Write a program to illustrate the use of the break statement.

```
public class Break
```

```java
{
    public static void main(String[] args)
    {
        int num = 0;
        while(num< 5)
        {
            if(num == 2)
            {
                System.out.println("Hello!!");
                break;
            }
            System.out.println("Number = " +num);
            num = num + 1;
        }
    }
}
```

The output of the program is shown as:

```
Administrator: C:\Windows\system32\cmd.exe

C:\Users\user\Desktop>javac Break.java

C:\Users\user\Desktop>java Break
Number = 0
Number = 1
Hello!!

C:\Users\user\Desktop>
```

In the previous code, when the value of num is equal to two, the break statement is executed and the control jumps out of the while loop following the next statement after the while loop. Hence, the break statement is used to exit from a loop at any point.

As we see a break statement is used to exit a particular loop, while a continue statement is used for doing the next iteration of the loop. Continue statements are also used with loops. Unlike with a break statement, the loop does not terminate when a continue statement is encountered. *A continue statement skips the rest of the statements, and the control is transferred to the loop-continuation portion of the loop.* Therefore, the execution of the loop resumes with the next iteration. The syntax of the continue statement is as follows:

Syntax –continue;

For example:

//Write a program to illustrate the use of the continue statement.

```java
public class Continue
{
    public static void main(String[] args)
    {
        int n;
        for(n = 0 ; n <= 8 ; n++)
        {
            if(n == 4)
            continue;
            System.out.println(n);
        }
    }
}
```

The output of the program is shown as:

In the previous code, as soon as the value of n becomes equal to four, the continue statement is executed, and the system.out.println() statement is skipped. The control is transferred to the expression, which increments the value of n. Hence, a *control* statement is the opposite of a *break* statement.

2.14   Methods in Java

As Java programmers, we often experience that the size of our program becomes too large and its complexity also increases. So, at that time it is very difficult for a programmer to read the entire code and also to check for any errors in it. Hence, to overcome this problem, *Java language enables us to break the entire program into a smaller number of modules or segments. These modules or segments are called methods. Therefore, a method is a predefined block of code designed to perform a particular task.* Methods are used to improve the efficiency of the program. Methods can reduce redundancy and help so that the code is easily understood. Each method is designed to perform a particular task. Methods are separated into two categories:

1. **Library Methods:** Library methods are those methods which are predefined.

2. **User-Defined Methods:** Unlike predefined library methods, these methods can be defined by the programmer or user. We can easily create these types of methods. The general form of a user-defined method is as follows:

```
[return type] <Method name> (parameters/ arguments)
{
    Statements;
    return;
}
```

Where:

(a) **Return Type:** Return types are used to identify which kind of value is going to be returned by the methods. Return types are the data types. If a method does not return any value, then the return type is void.

(b) **Method Name:** It identifies the name of a method. The name of the method should not be reserved in the Java libraries.

(c) **Parameters/Arguments:** These are the variables or values passed with their data types to the method for performing various operations.

(d) **Statements:** Statements are the particular steps that are performed by the method.

2.15   Summary

- The Java language was created and developed at Sun Microsystems, Inc. in 1991 by James Gosling, Patrick Naughton, Chris Warth, Ed Frank, and Mike Sheridan.

- The Java language supports all the features, flexibilities, and attributes of the C language. It also introduces various new features that were designed to support object-oriented programming. Java was initially known with the new name of "C with Classes." Thus, Java is the object-oriented version of C.

- An *object* is a container in which the data is stored in a combination of variables, methods, and data structures.

- A *class* is basically a user-defined data type that holds its own data members and member methods. It is a way to bind the data members and its methods together. A class can be accessed by creating an object of that class.

- *Encapsulation* is a process of wrapping up data members and member methods into a single unit. It is the mechanism which keeps the code and the data safe from external interference.

- *Inheritance* is the process of deriving a new class from the existing one. The existing class is known as the "base class," "parent class," or "superclass." The new class is known as the "derived class," "child class," or "subclass."

- The word *polymorphism* is derived from the word *poly*, which means many, and the word *morph*, which means forms. Polymorphism means the ability to make more than one form.

- There are two types of polymorphism, which are compile time and runtime polymorphism.

- *Abstraction* refers to the act of displaying only the essential features and hiding the background details and explanations which are not needed for presentation. This is also known as data abstraction.

- Keywords in Java are the reserved words which have a special meaning. They are written in lower case. Keywords cannot be used as an identifier. An identifier is a name which is given to a constant, variable, method, or array.

- Data types in Java are the special keywords which define the type of data and amount of data a variable is holding.

- The four basic data types used in Java are `int`, `char`, `float`, and `boolean`.

- Operators in Java are used to perform some specific operations between the different variables and constants.

- Arithmetic operators are those operators which are used in mathematical calculations.

- Assignment operators are used for assigning values to the variables. These operators are always evaluated from right to left.

- Relational operators are used for comparison between two values or expressions. They are also known as comparison operators.

- A conditional operator is also known as a ternary operator because it takes three operands.

- Bitwise operators are the special operators that are used to perform operations at the bit level.

- The comma operator is used to chain together some expressions.

- A unary operator is one which requires only a single operand to work. Java supports two unary operators, which are the increment (++) and decrement (--) operators. These operators are used to increase or decrease the value of a variable by one respectively.

- Control statements are those that enable a programmer to execute a particular block of code specifying the order in which the various instructions in a program are required to be executed. They determine the flow of control.

- The IF statement is a bidirectional control statement which is used to test the condition and take one of the possible actions.

- Nested IF-ELSE statements are also known as IF-ELSE-IF ladders.

- A switch statement is a multidirectional conditional control statement. It is a simplified version of an IF-ELSE-IF statement.

- Looping statements, also known as iterative statements, are sets of instructions which are repeatedly executed until a certain condition or expression becomes false.

▪ A while loop is a loop which is used to repeat a set of one or more instructions/statements until a particular condition becomes false.

▪ A do-while loop is similar to a while loop. The only difference is that, unlike a while loop in which a condition is checked at the start of the loop, in a do-while loop the condition is checked at the end of the loop.

▪ A for loop is a pre-deterministic loop; that is, it is a count-controlled loop such that the program knows in advance how many times the loop is to be executed.

▪ A break statement causes an intermediate exit from that loop in which the statement appears. A continue statement skips rest of the statements, and the control is transferred to the start of the loop.

▪ Java language enables us to break the entire program into a smaller number of modules or segments. These modules or segments are called methods. Therefore, a method is a predefined block of code designed to perform a particular task.

2.16 Exercises

2.16.1 Theory Questions

1. Explain the various characteristics of the Java programming language.

2. What is meant by the term inheritance? Discuss all the types in detail.

3. Define the term data abstraction. Give the difference between data hiding and data abstraction.

4. What is a class in Java?

5. Define polymorphism. What are the various types of polymorphism?

6. What are the different operators used in Java? Discuss all of them in detail.

7. What are the data types in Java? Explain in detail.

8. Explain the benefits of object-oriented programming (OOP).

9. What are the features of an object-oriented programming language?

10. What do you understand about Java tokens? Discuss in detail.

11. Define the term identifiers. What are various rules for identifying an identifier? Give examples.

12. What do you understand about the conditional operator? Explain with the help of an example.

13. What are decisional control statements in Java?

14. Differentiate between while and do-while loops. Give examples.

15. What is the difference between simple IF and IF-ELSE statements? Explain with the help of an example.

16. Write the syntax of a for loop. Can we skip any part in a for loop or not?

17. Make a class in which two members should be declared as private and three members should be declared as public.

18. Explain the switch case statement. What are the various advantages of using a switch case statement?

19. What do you understand about iterative statements in Java? Briefly discuss all the types.

20. Give the difference between break and continue statements with suitable examples.

21. Explain how objects help us in accessing the private data members of a class.

22. Differentiate between prefix and postfix increment operators.

2.16.2 Programming Questions

1. Write a Java program to print "Hello World" on the screen.

2. Write a program that reads five integer values and displays them.

3. Write a Java program to add two floating-point numbers.

4. Write a program to check whether the given number is even or odd.

5. Write a program to find the largest of three given numbers.

6. Write a menu-driven program performing addition, subtraction, multiplication, and division of two numbers.

7. Write a program where both IF and ELSE statements are executed in a program.

8. Write a program to find the factorial of a number using a for loop.

9. Write a program to accept a string from the user and to check whether the string is a palindrome or not.

10. Write a Java program to print the numbers from 1 to 10 excluding 5 using a continue statement.

11. Write a Java program to print a Fibonacci series up to 200.

2.16.3   Multiple Choice Questions

1. A(n) _____ is used to bind the data members and member methods together.
   (a) Object
   (b) Class
   (c) Array
   (d) None of the above

2. A conditional operator is also called a ternary operator as it has _____ operands.
   (a) 1
   (b) 3
   (c) 2
   (d) 4

3. The process by which different objects communicate with each other is called _____.
   (a) Message passing
   (b) Dynamic binding
   (c) Communication
   (d) Polymorphism

4. The process of deriving a new class from an existing class is called _____.
   (a) Polymorphism
   (b) Abstraction
   (c) Inheritance
   (d) Encapsulation

5. Which of the following is a valid identifier in Java?

(a) a_43

(b) cd bd

(c) apple

(d) both (a) and (c)

6. Which operator is used for mathematical computation?

(a) Assignment operator

(b) Arithmetic operator

(c) Bitwise operator

(d) Relational operator

7. Which operator is used for comparison between values?

(a) Logical operator

(b) Relational operator

(c) Assignment operator

(d) Unary operator

8. In which of the following loops will a block of statements be executed at least once without checking the condition?

(a) For loop

(b) While loop

(c) Do-while loop

(d) All of the above

9. The process of wrapping up of data members and member method into a single unit is known as _____.

(a) Polymorphism

(b) Abstraction

(c) Inheritance

(d) Encapsulation

CHAPTER 3

ARRAYS

3.1 Introduction

In the previous chapter we studied the basics of programming in data structures and Java in which we aimed to design good programs, where a good program refers to a program which runs correctly and efficiently by occupying less space in the memory, and also takes less time to run and execute. Undoubtedly, a program is said to be efficient when it executes with less memory space and also in minimal time. In this chapter, we will learn about the concept of arrays. An array is a user-defined data type that stores related information together. Arrays are discussed in detail in the following sections.

3.2 Definition of an Array

An array is a collection of homogeneous (similar) types of data elements in contiguous memory. An array is a linear data structure, because all elements of the array are stored in linear order. Let us take an example in which we have ten students in a class, and we have been asked to store the marks of all ten students; then we need a data structure known as an array.

36	98	14	74	56	13	7	96	44	82
marks1	marks2	marks3	marks4	marks5	marks6	marks7	marks8	marks9	marks10

Figure 3.1. Representation of an array of ten elements.

In the previous example, the data elements are stored in the successive memory locations and are identified by an index number (also known as the subscript), that is, A_i or A[i]. *A subscript is an ordinal number which is used to identify an element of the array.* The elements of an array have the same data type, and each element in an array can be accessed using the same name.

Frequently Asked Questions

1. What is an array? How can we identify an element in the array?

Ans: *An array is a collection of homogeneous (similar) types of data elements in contiguous memory. An element in an array can be identified by its index number, which is also known as a subscript.*

3.3 Array Declaration

We know that all variables must be declared before they are used in the program. Therefore, the same concept also holds with array variables. An array must be declared before it is used. During the declaration of an array, the size of the array has to be specified. Declaring an array involves the following specifications:

- *Data Type*: The data type means the different kinds of values it can store. The data type can be an `integer (int)`, `float`, `char`, or any other valid data type.

- *Array Name*: The name refers to the name of the array which will be used to identify the array.

- *Size*: The size of an array refers to the maximum number of values an array can hold.

Syntax: `data_type [] array_name = new  data_type[size] ;`

`Example: int [] salary = new int[10];`

The previous example declares salary to be an array which has ten elements. In Java, the array index starts from zero. The first element of this array will be stored in salary [0], the second element will be stored in salary [1], and so on. Similarly, the last element will be stored in salary [9]. In memory, the array will be shown as in the following figure.

2000	4500	7890	9876	10000	3458	8000	9810	14000	5000
salary[0]	salary[1]	salary[2]	salary[3]	salary[4]	salary[5]	salary[6]	salary[7]	salary[8]	salary[9]

Figure 3.2. Memory representation of an array.

Here 0, 1, 2, . . . 9 written in square brackets represent the subscripts which we use to identify a particular element in the array.

3.4 Array Initialization

The initialization of arrays can be done in the following ways:

1. **Initialization at Compile Time:** Initialization of elements of the array at compile time refers to the same way we initialize the normal or ordinary variables at the time of their declaration. When an array is initialized, there is a need to provide a specific value for every element in the array.

   The general form of initializing arrays is as follows:

   ```
   data_type [] array_name = {list of values};
   ```

   During the initialization of arrays, we may omit the size of the array. For example,

   ```
   int [] age = {20, 25, 23, 28, 30};
   ```

   In the previous example, the compiler will automatically allocate memory for all the initialized elements of the array. If the number of values is less than the size provided, then such elements will take zeroes as their assigned values. For example,

   ```
   int [] marks;
   marks = new int [] {56, 69, 40,99, 82, 96, 72};
   ```

   Here the size of the array is 10, but there are only seven elements; hence, the remaining elements will be considered to be zeroes.

56	69	40	99	82	96	72	0	0	0
marks[0]	marks[1]	marks[2]	marks[3]	marks[4]	marks[5]	marks[6]	marks[7]	marks[8]	marks[9]

Figure 3.3. Initialization of array marks [10].

2. **Initialization at Runtime:** Initialization of elements of the array at runtime refers to the method of inputting the values from the keyboard. In this method, a while, do-while, or for loop is taken to input the values of the array. The scanner class is used for taking input from the user. For using the object of the scanner class, first import the scanner class from java.util library and then create its instance to use the in-built functions.

```
import java.util.Scanner;  //Importing scanner class for input
public class ArrayDemo {
    public static void main(String[] args) {
        Scanner scn = new Scanner(System.in);
        int[] salary = new int[15];
        for (int i = 0; i < salary.length; i++) {
            salary[i] = scn.nextInt(); //Inputting the values
        }   //End of for
    } //End of main
}   //End of class
```

Figure 3.4. Code for inputting the values.

In the previous code, the index i is at 0, and the values will be input for the index values from 0 to 14, as the array has 15 elements.

3.5   Calculating the Address of Array Elements

The address of the elements in the 1-D array can be calculated very easily, because the array stores all its data elements in contiguous memory locations, storing the base address (address of the first element of the array). Hence, the address of the other data elements can easily be calculated using the base address. The formula to find the address of elements in a 1-D array is as follows:

Address of data element, A[i] = Base Address (BA) + w (i − lower bound)

where A is the array, i is the index of the element for which the address is to be calculated, BA is the base address of the array A, and w is the size of each element (e.g., the size of int is 2 bytes, the size of char is 1 byte, etc.)

Frequently Asked Questions

2. An array is given int marks [6] = {34, 53, 87, 100, 98, and 65}; calculate the address of marks [3] if the base address is 3000.

Ans: *It is given that the base address of the array is 3000, and we know that the size of an integer is 2 bytes. Hence, we can easily find the address of marks [3].*

3000	3002	3004	3006	3008	3010
34	53	87	100	98	65
marks[1]	marks[2]	marks[3]	marks[4]	marks[5]	marks[6]

> By putting into the formula –
> Address of marks[3] = 3000 + 2 (3 – 1)
>   = 3000 + 2 (2)
> Address of marks[3] = 3004

3.6 Operations on Arrays

This section discusses various operations that can be performed on arrays. These operations include:

- Traversing an array
- Inserting an element in an array
- Deleting an element in an array
- Searching an element in an array
- Merging of two arrays
- Sorting an array

1. Traversing an Array

Traversing an array means to access every element in an array exactly once so that it can be processed. Examples are counting all the data elements, performing any process on these elements, and so on. Traversing the elements of the array is a very simple process because of the linear structure of the array (all the elements are stored in contiguous memory locations).

Practical Application

If there is a line of people standing one after the other, and one boy is distributing advertisement pamphlets one by one to each person standing in the line.

```java
import java.util.Scanner; //Importing scanner class for input

public class ArrayDemo {
    public static void main(String[] args) {

        Scanner scn = new Scanner(System.in);
        int[] num = new int[5];
        System.out.println("Enter the elements of array: ");
        for (int i = 0; i < num.length; i++) {
                num[i] = scn.nextInt(); // Inputting the values
```

```
        } // End of loop1

        System.out.println("The elements of array are: ");
        for (int i = 0; i < num.length; i++) {
            System.out.print(num[i] + "\t");
        } // End of loop2
    } // End of main
} // End of class
```

The output of the program is shown as:

In the previous code, the traversing of elements of the array is shown. In the first for loop, all the elements are inputted into the array. Second, all the elements are traversed and counted in the second for loop. Hence, the traversing of an array is done.

2. Inserting an Element in an Array

Inserting an element in an array refers to the operation of adding an element to the array. In the case of insertion, we assume that there is enough memory space still available in the array. For example, if we have an array that can hold twenty elements and the array contains only fifteen elements, then we have space to accommodate five more elements. However, if the array can hold fifteen elements, then we will not be able to insert other elements into the array. Insertion in arrays can be done in three ways:

(a) Insertion at the beginning

(b) Insertion at a specified position

(c) Insertion at the end

Now let us discuss all of these cases in detail.

(a) Insertion at the beginning: In this case, the new element to be inserted is inserted at the beginning of the array. To insert an element at the beginning, all the elements stored in the array must move one place forward to vacate the first position in the array. For example, if an array is declared to hold ten elements and it contains only seven elements, and also if it is given that the new element is to be inserted at the beginning of the array, then all the stored elements must move one place ahead, which is shown as follows:

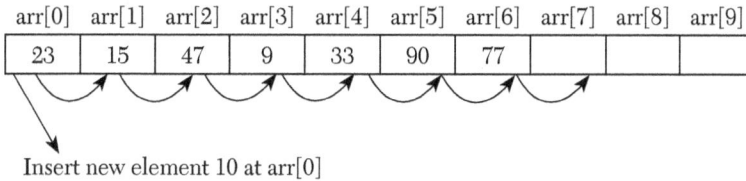

arr[0]	arr[1]	arr[2]	arr[3]	arr[4]	arr[5]	arr[6]	arr[7]	arr[8]	arr[9]
23	15	47	9	33	90	77			

Insert new element 10 at arr[0]

After shifting all the existing elements toward the right and inserting new element 10 into the first slot of the array, the new array will be:

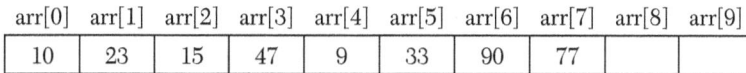

arr[0]	arr[1]	arr[2]	arr[3]	arr[4]	arr[5]	arr[6]	arr[7]	arr[8]	arr[9]
10	23	15	47	9	33	90	77		

Algorithm for Insertion in the Beginning

We assume *ARR* is an array with *N* elements in it. The maximum elements that can be stored in the array is defined by *SIZE*. We should first check if the array has an empty space available to store any element in it or not, and then we proceed with the insertion process.

```
Step 1: START
Step 2: IF N = SIZE,
          Print OVERFLOW
          ELSE
          N = N + 1
Step 3: SET I = N
Step 4: Repeat Step 5 while I>=0
Step 5: SET ARR[I+1] = ARR[I]
          [ END OF LOOP ]
Step 6: SET ARR[0] = New_Element
Step 7: EXIT
```

(b) Insertion at a specified position: In this case the new element to be inserted is inserted at a specified location/position which is entered by the user. In order to insert a new element in the array, the previously stored elements in the array must move one place forward from their current place until the element at the specified position is reached. For example, if an array is declared to hold ten elements and it contains only eight elements, and it is also given that the new element is to be inserted at the fifth position of the array, then the stored elements must move one place ahead as shown in the following:

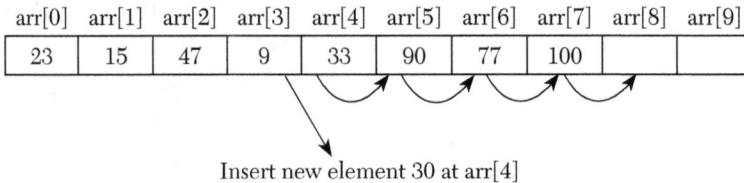

arr[0]	arr[1]	arr[2]	arr[3]	arr[4]	arr[5]	arr[6]	arr[7]	arr[8]	arr[9]
23	15	47	9	33	90	77	100		

Insert new element 30 at arr[4]

After shifting the elements from the middle position and inserting new element 30 into the middle of the array, the new array will be:

arr[0]	arr[1]	arr[2]	arr[3]	arr[4]	arr[5]	arr[6]	arr[7]	arr[8]	arr[9]
23	15	47	9	30	33	90	77	100	

Practical Application

It is just like if there are people standing in a line and one person just joins the line from the middle, so now every person has to shift one place backward from the middle so that the person can come in the line; hence, it is insertion at the middle.

Algorithm for Insertion at a Specified Position

We assume *ARR* is an array with *N* elements in it. The maximum elements that can be stored in the array is defined by *SIZE*. Let *POS* define the position at which the new element is to be inserted. We should first check if the array has an empty space available to store any element in it or not, and then we proceed with the insertion process.

```
Step 1:START
Step 2:IF N = SIZE,
        Print OVERFLOW
        ELSE
        N = N + 1
Step 3:SET I = N
Step 4:Repeat Step 5 while I>=POS
Step 5:SET ARR[I+1] = ARR[I]
        [ END OF LOOP ]
```

```
Step 6:SET ARR[POS] = New_Element
Step 7:EXIT
```

(c) Insertion at the end: In this case the new element to be inserted is inserted at the end of the array. So, there is no need for swapping the elements in this case. We are just required to check whether there is enough space available in the array or not. For example, if an array is declared to hold ten elements and it contains only nine elements, then the insertion can take place.

arr[0]	arr[1]	arr[2]	arr[3]	arr[4]	arr[5]	arr[6]	arr[7]	arr[8]	arr[9]
23	15	47	9	30	33	90	77	100	

Insert new element 11 at arr[9]

Now the last element will be inserted at the last position, which is at arr [9] and is vacant. Therefore, the new array after insertion will be:

arr[0]	arr[1]	arr[2]	arr[3]	arr[4]	arr[5]	arr[6]	arr[7]	arr[8]	arr[9]
23	15	47	9	30	33	90	77	100	11

Practical Application

It is just like a normal line where a person comes and joins the line at the end; hence, there is no need for any shifting in this process.

Algorithm for Insertion at the End

We assume *ARR* is an array with *N* elements in it. The maximum elements that can be stored in the array is defined by *SIZE*. We should first check if the array has an empty space available to store any element in it or not, and then we proceed with the insertion process.

```
Step 1:START
Step 2:IF N = SIZE,
        Print OVERFLOW
        ELSE
        N = N + 1
Step 3:SET ARR[N] = New_Element
Step 4:EXIT
```

// Write a menu-driven program to implement insertion in a 1-D array discussing all three cases.

```java
import java.util.Scanner;
public class ArrayInsertionDemo {

    public static void main(String[] args) {
        Scanner scn = new Scanner(System.in);
        int[] arr = new int[10];
        System.out.println("Enter the number of elements
                            in array:");
        int n = scn.nextInt();
        System.out.println("Enter the elements of array:");
        for (int i = 0; i < n; i++) {
            arr[i] = scn.nextInt();
        }

        System.out.println("***MENU***");
        System.out.println("1. Insertion from beginning");
        System.out.println("2. Insertion from specified location");
        System.out.println("3. Insertion from end");
        System.out.println("Enter your choice:");
        int choice = scn.nextInt();
        int value = 0;

        if (n == 10) {
            System.out.println("Overflow error");
        } else {
            switch (choice) {
            case 1:
                for (int i = n - 1; i >= 0; i--) {
                    arr[i + 1] = arr[i];
                }
                System.out.println("Enter new value:");
                value = scn.nextInt();
                arr[0] = value;
                System.out.println("After insertion array is");
                for (int i = 0; i <= n; i++) {
                    System.out.print(arr[i] + "\t");
                }
                break;

            case 2:
                System.out.println("Enter position");
                int pos = scn.nextInt();
                for (int i = n - 1; i >= pos - 1; i--) {
                    arr[i + 1] = arr[i];
```

```
            }
            System.out.println("Enter new value:");
            value = scn.nextInt();
            arr[pos - 1] = value;
            System.out.println("After insertion array is");
            for (int i = 0; i <= n; i++) {
                System.out.print(arr[i] + "\t");
            }
            break;
        case 3:
            System.out.println("Enter new value:");
            value = scn.nextInt();
            arr[n] = value;
            System.out.println("After insertion array is");
            for (int i = 0; i <= n; i++) {
                System.out.print(arr[i] + "\t");
            }
            break;
        default:
            System.out.println("Wrong Choice");
            break;
        }
    }
}
}
```

The output of the program is shown as:

3. Deleting an Element in an Array

Deleting an element from an array refers to the operation of the removal of an element from an array. Deletion in an array can be done in three ways:

(a) Deletion from the beginning

(b) Deletion from a specified position

(c) Deletion from the end

Now let us discuss all of these cases in detail.

(a) Deletion from the beginning: In this case the element to be deleted is deleted from the beginning of the array. In order to delete an element from the beginning, all the elements stored in the array must move one place backward in the array. For example, if an array is declared to hold ten elements and it contains only seven elements, and it is also given that the element is to be deleted from the beginning of the array, then all the stored elements must move one place back as shown in the following array:

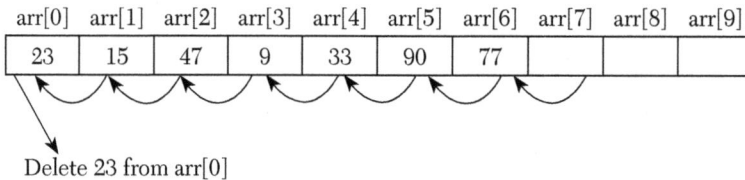

Delete 23 from arr[0]

In order to delete the first element, 23, from the array, we must swap all the stored elements backward so that the first element gets deleted as shown. After the deletion of 23 from the array, the new array will be:

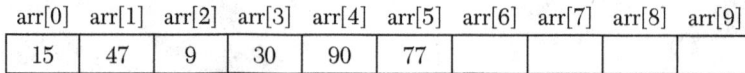

arr[0]	arr[1]	arr[2]	arr[3]	arr[4]	arr[5]	arr[6]	arr[7]	arr[8]	arr[9]
15	47	9	30	90	77				

Practical Application

It is just like if there is a pile of books and a person just picks up the book from the end, so now all the books will be shifted one place forward from where they were placed; hence, this is deletion from the beginning.

Algorithm for Deletion from the Beginning

We assume *ARR* is an array with *N* elements in it.

```
Step 1: START
Step 2: SET I = 0
Step 3: Repeat Step 4 while I<N-1
```

```
Step 4:SET ARR[I] = ARR[I+1]
        [ END OF LOOP ]
Step 5:EXIT
```

(b) Deletion from a specified position: In this case the element to be deleted is deleted from the specified location/position in the array which is entered by the user. In order to delete an element from the specified position, the elements stored in the array must move one place backward to their existing place in the array until the element is deleted at the specified position. For example, if an array is declared to hold ten elements and it only contains eight elements, and it is also given that the element is to be deleted from the specified position which is the fourth position of the array, then the stored elements must move one place back as shown in the following:

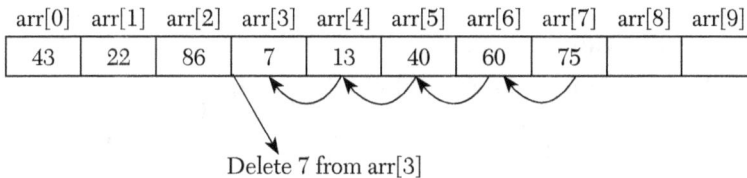

arr[0]	arr[1]	arr[2]	arr[3]	arr[4]	arr[5]	arr[6]	arr[7]	arr[8]	arr[9]
43	22	86	7	13	40	60	75		

Delete 7 from arr[3]

In order to delete the fourth element, 7, from the array, we must swap the stored elements backward so that the given element gets deleted as shown. After the deletion of 7 from the array, the new array will be:

arr[0]	arr[1]	arr[2]	arr[3]	arr[4]	arr[5]	arr[6]	arr[7]	arr[8]	arr[9]
43	22	86	13	40	60	75			

Algorithm for Deletion from a Specified Position

We assume *ARR* is an array with *N* elements in it. Let *POS* define the position from which the element is to be deleted.

```
Step 1:START
Step 2:SET I = POS
Step 3:Repeat Step 4 while I<=N-1
Step 4:SET ARR[I+1] = ARR[I]
        [ END OF LOOP ]
Step 5:EXIT
```

(c) Deletion from the end: In this case the deletion is quite simple. Here we are just required to count all the elements except the last one, as we want to delete the last element. For example, if an array is declared to hold ten elements and it contains only six elements, and it is also given that the element is to be deleted from the end of the array, then the deletion is shown as follows:

arr[0]	arr[1]	arr[2]	arr[3]	arr[4]	arr[5]	arr[6]	arr[7]	arr[8]	arr[9]
43	22	86	13	40	11				

Delete 11 from arr[5]

After deleting 11 from the array, the new array will be:

arr[0]	arr[1]	arr[2]	arr[3]	arr[4]	arr[5]	arr[6]	arr[7]	arr[8]	arr[9]
43	22	86	13	40					

Practical Application

It is just like if there is a pile of books and a person just picks the first book from the pile; hence, we can say that one book is deleted or removed from the pile, and therefore it is deletion from the end.

Algorithm for Deletion from the End

We assume *ARR* is an array with *N* elements in it.

```
Step 1:START
Step 2:SET N = N-1
Step 3:Repeat Step 4 for I=0 to N
Step 4:Display ARR[I]
       [ END OF LOOP ]
Step 5:EXIT
```

// Write a menu-driven program to implement deletion in a 1-D array discussing all three cases.

```java
import java.util.Scanner;
public class ArrayDeletionDemo {
    public static void main(String[] args) {
        Scanner scn = new Scanner(System.in);
        int[] arr = new int[10];
        System.out.println("Enter the number of elements
                            in array:");
        int n = scn.nextInt();
        System.out.println("Enter the elements of array:");
        for (int i = 0; i < n; i++) {
            arr[i] = scn.nextInt();
```

```
}
System.out.println("***MENU***");
System.out.println("1. Deletion from beginning");
System.out.println("2. Deletion from specified location");
System.out.println("3. Deletion from end");
System.out.println("Enter your choice:");
int choice = scn.nextInt();

if (n == 10) {
    System.out.println("Overflow error");

} else {
    switch (choice) {

    case 1:
        for (int i = 0; i < n - 1; i++) {
            arr[i] = arr[i + 1];
        }
        System.out.println(" After deletion array is");
        for (int i = 0; i < n - 1; i++) {
            System.out.print(arr[i] + "\t");
        }
        break;

    case 2:
        System.out.println(" Enter position");
        int pos = scn.nextInt();
        for (int i = pos - 1; i < n - 1; i++) {
            arr[i] = arr[i + 1];
        }
        System.out.println(" After deletion array is");
        for (int i = 0; i < n - 1; i++) {
            System.out.print(arr[i] + "\t");
        }
        break;

    case 3:
        n = n - 1;
        System.out.println(" After deletion array is");
        for (int i = 0; i < n - 1; i++) {
            System.out.print(arr[i] + "\t");
        }
        break;
```

```
            default:
                System.out.println("Wrong Choice");
                break;
            }
        }
    }
}
```

The output of the program is shown as:

4. Searching for an Element in an Array

Searching for an element in an array means to find whether a particular value exists in an array or not. If that particular value is found, then the searching is said to be successful and the position/location of that particular value is returned. If the value is not found, then searching will be said to be unsuccessful. There are two methods for searching, linear search and binary search. In this chapter we will only discuss linear search in detail, and the binary search technique will be discussed in upcoming chapters. Now, we will learn how linear search works.

Linear Search

Linear search is a very simple technique used to search for a particular value in an array. It is also called a sequential search, as it works by comparing the values to be searched with every element of the array in a sequence until a match is found.

For example, let us take an array of ten elements which is declared as:

```
int [] array = new int [10];
    array = {23, 15, 47, 9, 30, 33, 90, 77, 100, 11}
```

and search for 90 in the array; then, every element of the array will be compared to 90 until 90 is found.

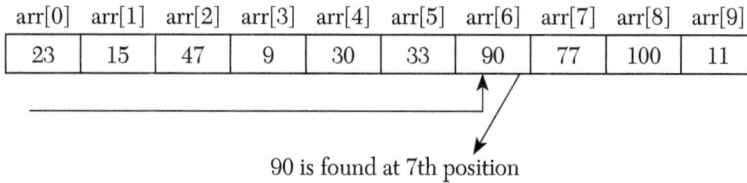

arr[0]	arr[1]	arr[2]	arr[3]	arr[4]	arr[5]	arr[6]	arr[7]	arr[8]	arr[9]
23	15	47	9	30	33	90	77	100	11

90 is found at 7th position

In this way a linear search is used to search for a particular value in the array. The following is the program for a linear search.

//Write a program to search for an element in an array using the linear search technique.

```java
import java.util.Scanner;
public class LinearSearchDemo {

    public static void main(String[] args) {
        Scanner scn = new Scanner(System.in);

        int[] arr;
        System.out.println("***LINEAR SEARCH***");
        System.out.println("Enter number of elements in array:");
        int n = scn.nextInt();
        arr = new int[n];
        System.out.println("Enter elements of array:");
        for (int i = 0; i < n; i++) {
            arr[i] = scn.nextInt();

        }
        System.out.println("Enter value to search:");
        int value = scn.nextInt();
        int r = linear_search(arr, n, value);
        if (r == -1)
            System.out.println(" Value not found");
        else

            System.out.println(value + " found at " + (r + 1) +
                            " position");

    }
    public static int linear_search(int arr[], int n, int value) {
```

```
int i;
for (i = 0; i < n; i++) {
    if (arr[i] == value)
        return i;

}
return (-1);
}


}
```

The output of the program is shown as:

5. Merging of Two Arrays

The merging of two arrays means copying the elements of the first and second array into the third array. Here we will take two sorted arrays, and the resultant merged array will also be sorted. The concept of merging is explained as follows:

Let us consider two sorted arrays, Array 1 and Array 2, and an Array 3 in which the elements will be placed after sorting.

Array: 1

arr[0]	arr[1]	arr[2]	arr[3]	arr[4]	arr[5]	arr[6]	arr[7]	arr[8]	arr[9]
6	12	18	24	30					

Array: 2

arr[0]	arr[1]	arr[2]	arr[3]	arr[4]	arr[5]	arr[6]	arr[7]	arr[8]	arr[9]
7	14	21	28	35					

Array: 3

arr[0]	arr[1]	arr[2]	arr[3]	arr[4]	arr[5]	arr[6]	arr[7]	arr[8]	arr[9]
6	7	12	14	18	21	24	28	30	35

Array 3 shows how the merged array is formed using the sorted Arrays 1 and 2. Here we compare the elements of the two arrays. First, the first element of Array 1 is compared with the first element of Array 2, and as 6 is less than 7 (6 < 7), therefore 6 will be the first element in the merged array. Now the second element of Array 1 is compared to the first element of Array 2, and as 7 is less than 12 (7 < 12), therefore 7 will be the second element in the merged array. Now the second element of Array 1 is compared with the second element of Array 2, and as 12 is less than 14 (12 < 14), therefore 12 will be the third element in Array 3. This procedure is repeated until the elements of both Arrays 1 and 2 are placed in the right positions in the merged array, that is, Array 3.

Practical Application

A real life example of merging would be if there are two different lines and both lines need to be merged according to the height of the people standing in that line, and then merging would be done into a new line where the new line would consist of people from both lines in which people would be standing in order according to their heights.

// Write a program to merge two sorted arrays.

```java
import java.util.Scanner;
public class MergeSortedArraysDemo {

    public static void main(String[] args) {
        Scanner scn = new Scanner(System.in);
        int[] arr1, arr2, arr3;
        int i = 0, j = 0, k = 0;
        System.out.println("Enter number of elements in array 1:");
        int m1 = scn.nextInt();
        arr1 = new int[m1];
        System.out.println("Enter elements of array 1:");
        for (int s = 0; s < m1; s++) {
            arr1[s] = scn.nextInt();
        }

        System.out.println("Enter number of elements in array 2:");
        int m2 = scn.nextInt();
```

```java
    arr2 = new int[m2];
    System.out.println("Enter elements of array 2:");
    for (int s = 0; s < m2; s++) {
        arr2[s] = scn.nextInt();
    }

    arr3 = new int[m1 + m2];
    while (i < m1 && j < m2) {
        if (arr1[i] < arr2[j]) {
            arr3[k] = arr1[i];
            k++;
            i++;
        } else if (arr2[j] < arr1[i]) {
            arr3[k] = arr2[j];
            j++;
            k++;

        } else {
            arr3[k] = arr1[i];
            k++;
            i++;
            arr3[k] = arr2[j];
            k++;
            j++;
        }

    } // End of while loop

    while (i < m1) {
        arr3[k] = arr1[i];
        k++;
        i++;
    }
    while (j < m2) {
        arr3[k] = arr2[j];
        k++;
        j++;
    }
    System.out.println("After merging new array is:");
    for (int s = 0; s < (m1 + m2); s++) {
        System.out.print(arr3[s] + " ");
    }
    }
}
```

The output of the program is shown as:

6. Sorting an Array

Sorting an array means arranging the data elements of a data structure in a specified order either in ascending or descending order. Sorting refers to the process where, for example, in a class of sixty students who have gotten grades on their examination, now the names of the students will be counted according to their grades either in ascending or descending order.

For example: If we have an array of ten elements declared as

int [] array = {78, 12, 47, 55, 61, 6, 99, 84, 32, 10}

then after sorting, the new array will be:

array = {6, 10, 12, 32, 47, 55, 61, 78, 84, 99}

There are various types of sorting techniques which include *selection sort*, *insertion sort*, and *merge sort*; we will learn about selection sort in this chapter, and the other techniques will be discussed in an upcoming chapter.

Selection Sort

Selection sort is a sorting technique that works by finding the smallest value in the array and placing it in the first position. After that, it then finds the second smallest value and places it in the second position. This process is repeated until the whole array is sorted. It is a very simple technique, and it is also easier to implement than any other sorting technique. Selection sort is generally used for sorting large records.

Selection Sort Technique

Consider an array with N elements.

Pass 1: Find the position POS of smallest value in the array of N elements and interchange ARR [POS] with ARR [0]. Thus, A [0] is sorted.

Pass 2: Find the position POS of smallest value in the array of N-1 elements and interchange ARR [POS] with A [1]. Thus, A [1] is sorted.

⋮

Pass N-1: Find the position POS of the smaller of the elements ARR [N-2] and ARR [N-1] and interchange ARR [POS] with ARR [N-2]. Thus, ARR [0], ARR [1], ARR [N-1] is sorted.

For example: Sort the given array using selection sort.

arr[0]	arr[1]	arr[2]	arr[3]	arr[4]	arr[5]	arr[6]	arr[7]	arr[8]	arr[9]
16	82	48	34	75	9				

Pass	POS	Array[0]	Array[1]	Array[2]	Array[3]	Array[4]	Array[5]
1	5	9	82	48	34	75	16
2	5	9	16	48	34	75	82
3	3	9	16	34	48	75	82
4	3	9	16	34	48	75	82
5	4	9	16	34	48	75	82
6	5	9	16	34	48	75	82

This is the way the selection sort technique works. The following is the program given for selection sort.

//Write a program to sort an array using the selection sort technique.

```java
import java.util.Scanner;
public class SelectionSortDemo {
    public static void main(String[] args) {
        Scanner scn = new Scanner(System.in);
        int[] arr;
        int min = 0;
        int pos = 0;
        int temp = 0;

        System.out.println("Enter number of elements in array:");
        int n = scn.nextInt();
        arr = new int[n];
        System.out.println("Enter elements of array:");
        for (int i = 0; i < n; i++) {
            arr[i] = scn.nextInt();
        }
        System.out.println("***Selection Sort***");
```

```java
for (int i = 1; i < n; i++) {
    min = arr[i - 1];
    pos = i - 1;
    for (int j = i; j < n; j++) {
        if (arr[j] < min) {
            min = arr[j];
            pos = j;
        }
    }
    if (pos != i - 1) {
        temp = arr[pos];
        arr[pos] = arr[i - 1];
        arr[i - 1] = temp;
    }

}

System.out.println("After sorting new array is:");
for (int i = 0; i < n; i++) {
    System.out.print(arr[i] + " ");
}
}

}
```

The output of the program is shown as:

3.7 2-D Arrays/Two-Dimensional Arrays

We have already discussed 1-D arrays/one-dimensional arrays and their various types and operations. Now, we will learn about two-dimensional arrays. Unlike one-dimensional arrays, 2-D arrays are organized in the form of grids or tables. They are a collection of 1-D arrays. One-dimensional arrays are linearly organized in the memory. A 2-D array consists of two subscripts:

1. first subscript: which denotes the row

2. second subscript: which denotes the column

A 2-D array is represented as shown in the following figure:

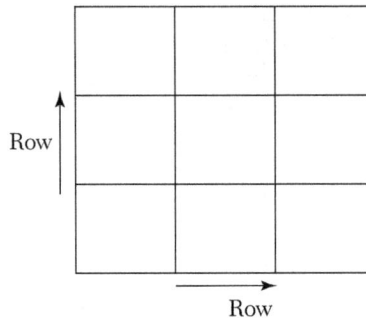

Figure 3.5. Representation of a 2-D array.

3.8   Declaration of Two-Dimensional Arrays

As we declared 1-D arrays, similarly we can declare two-dimensional arrays. For declaring two-dimensional arrays, we must know the name of the array, the data type of each element, and the size of each dimension (size of rows and columns).

Syntax: `data_type[][] array_name = new int [row_size]`
`        [column_size];`

A two-dimensional array is also called an m X n array, as it contains m X n elements where each element in the array can be accessed by i and j, where i<=m and j<=n, and where i, j, m, n are defined as follows:

i, j = subscripts of array elements,

m = number of rows,

n = number of columns.

For example: Let us take an array of 3 X 3 elements. Therefore, the array is declared as:

```
int marks [3] [3];
```

	0	1	2
0	marks[0][0]	marks[0][1]	marks[0][2]
1	marks[1][0]	marks[1][1]	marks[1][2]
2	marks[2][0]	marks[2][1]	marks[2][2]

Row

Row

In the previous diagram the array has three rows and three columns. The first element in the array is denoted by marks [0] [0]. Similarly, the second element will be denoted by marks [0] [1], and so on. Also, data elements in an array can be stored in the memory in two ways:

Row Major Order

In row major order the elements of the first row are stored before the elements of the second, third, and n rows. Here the data elements are stored on a row-by-row basis:

00	01	02	10	11	12	20	21	22					

Column Major Order

In column major order the elements of the first column are stored before the elements of the second, third, and n columns. Here the data elements are stored on a column-by-column basis:

00	10	20	01	11	21	02	12	22					

Now, we will calculate the base address of elements in a 2-D array, as the computer does not store the address of each element. It just stores the address of the first element and calculates the addresses of other elements from the base address of the first element of the array. Hence, the addresses of other elements can be calculated from the given base address.

1. Elements in Row Major Order

Address(A[i][j]) = Base address(BA) + w(n(i-1) + (j-1))

2. Elements in Column Major Order

Address(A[i][j]) = Base address(BA) + w (m(j-1) + (i-1)) where w is the size in bytes to store one element.

Frequently Asked Questions

3. Consider a 25 X 5 two-dimensional array of students which has a base address 500 and where the size of each element is 2. Now calculate the address of the element student[15][3], assuming that the elements are stored in

(a) Row major order
(b) Column major order
Ans:

(a) Row major order
Here we are given that $w = 2$, base address = 500, $n = 5$, $i = 15$, $j = 3$.
Address(A[i][j]) = Base address(BA) + w (n(i-1) + (j-1))
Address(student[15][3]) = 500 + 2 (5(15-1) + (3-1))
$$= 500 + 2 (5(14) + 2)$$
$$= 500 + 2 (72)$$
$$= 500 + 144$$
Address(student[15][3]) = 644

(b) Column major order
Here we are given that $w = 2$, base address = 500, $m = 25$, $i = 15$, $j = 3$
Address(A[i][j]) = Base address(BA) + w (m(j-1) + (i-1))
Address(student[15][3]) = 500 + 2 (25(3-1) + (15-1))
$$= 500 + 2 (25(2) + 14)$$
$$= 500 + 2 (64)$$
Address(student[15][3]) = 500 + 128 = 628

3.9 Operations on 2-D Arrays

There are various operations that are performed on two-dimensional arrays, which include:

- **Sum**: Let A_{ij} and B_{ij} be the two matrices which are to be added together, storing the result into the third matrix C_{ij}. Two matrices will be added when they are compatible with each other; that is, they should have the same number of rows and columns.

$$C_{ij} = A_{ij} + B_{ij}$$

- **Difference**: Let A_{ij} and B_{ij} be the two matrices which are to be subtracted together, storing the result into a third matrix C_{ij}. Two matrices will be subtracted when they are compatible with each other; that is, they should have the same number of rows and columns.

$$C_{ij} = A_{ij} - B_{ij}$$

- **Product**: Let A_{ij} and B_{ij} be the two matrices which are to be multiplied together, storing the result into a third matrix C_{ij}. Two matrices will be multiplied with each other if the number of columns in the first matrix is equal to the number of rows in the second matrix. Therefore, m X n matrix A can be multiplied with a p X q matrix B if n=p.

$$C_{ij} = A_{ik} \text{ X } B_{kj} \text{ for k=1 to n}$$

- **Transpose**: The transpose of an m X n matrix A is equal to an n X m matrix B, where

$$B_{ij} = A_{ij.}$$

// Write a program to read and display a 3 X 3 array.

```
import java.util.Scanner;

public class TwoDimensionalArrayDemo {

    public static void main(String[] args) {
        Scanner scn = new Scanner(System.in);
        int[][] array = new int[3][3];
        System.out.println("Enter the elements of array:");
        for (int i = 0; i < 3; i++) {
            for (int j = 0; j < 3; j++) {
                array[i][j] = scn.nextInt();
            }

        }
        System.out.println("The array is:");
        for (int i = 0; i < 3; i++) {
            System.out.println();
            for (int j = 0; j < 3; j++) {
                System.out.println("array[" + i + "]" +
                    "[" + j + "]" + " = " + array[i][j]);
            }

        }

    }

}
```

The output of the program is shown as:

// Write a program to find the sum of two matrices.

```java
import java.util.Scanner;
public class SumOfMatricesDemo {

    public static void main(String[] args) {
        Scanner scn = new Scanner(System.in);
        int[][] A = new int[2][2];
        int[][] B = new int[2][2];
        int[][] C = new int[2][2];
        System.out.println("Enter elements of array A:");
        for (int i = 0; i < 2; i++) {
            for (int j = 0; j < 2; j++) {
                A[i][j] = scn.nextInt();
            }
        }
        System.out.println("Enter elements of array B:");
        for (int i = 0; i < 2; i++) {
            for (int j = 0; j < 2; j++) {
                B[i][j] = scn.nextInt();
            }
        }
        for (int i = 0; i < 2; i++) {
            for (int j = 0; j < 2; j++) {
                C[i][j] = A[i][j] + B[i][j];
            }
        }
        System.out.println("Resultant matrix is :");
        for (int i = 0; i < 2; i++) {
```

```
        for (int j = 0; j < 2; j++) {
                System.out.println(C[i][j]);
            }


        }
    }
}
```

The output of the program is shown as:

```
// Write a program to find the transpose of a 3 X 3 matrix.
import java.util.Scanner;
public class TransposeDemo {

    public static void main(String[] args) {

        Scanner scn = new Scanner(System.in);
        int[][] matrix = new int[3][3];
        int[][] transpose = new int[3][3];

        System.out.println("Enter elements of matrix:");
        for (int i = 0; i < 3; i++) {
            for (int j = 0; j < 3; j++) {
                matrix[i][j] = scn.nextInt();
            }
        }
        System.out.println("The elements of matrix are:");
        for (int i = 0; i < 3; i++) {
            System.out.println();
            for (int j = 0; j < 3; j++) {
                System.out.print(matrix[i][j] + " ");
```

```
            }
        }
        for (int i = 0; i < 3; i++) {
            for (int j = 0; j < 3; j++) {
                transpose[i][j] = matrix[j][i];
            }
        }
        System.out.println("\n\nElements of transposed
                            matrix are:");
        for (int i = 0; i < 3; i++) {
            System.out.println();
            for (int j = 0; j < 3; j++) {
                System.out.print(transpose[i][j] + " ");
            }
        }
    }
}
```

The output of the program is shown as:

3.10 Multidimensional Arrays/N-Dimensional Arrays

A multidimensional array is also known as an n-dimensional array. It is an array of arrays. It has n indices in it which also justifies its name of n-dimensional array. An n-dimensional array is an $m_1 \times m_2 \times m_3 \times \ldots \times m_n$ array, as it contains $m_1 \times m_2 \times m_3 \times \ldots \times m_n$ elements. Multidimensional arrays are declared and initialized in the same way as one-dimensional and two-dimensional arrays.

3.11   Calculating the Address of 3-D Arrays

Just like 2-D arrays we can store 3-D arrays in two ways, row major order and column major order.

1. Elements in Row Major Order

Address ([i] [j] [k]) = Base Address (BA) + w (L_3 (L_2(E_1) + E_2) + E_3)

2. Elements in Column Major Order

Address ([i] [j] [k]) = Base Address (BA) + w (($E_3 L_2$ + E_2) L1 + E_1)

Where L is length of index, L = Upper bound – Lower bound + 1,

E is effective address, E = i – Lower bound.

Frequently Asked Questions

4. Let us take a 3-D array A (4:12, -2:1, 8:14) and calculate the address of A (5, 4, 9) using row major order and column major order, where the base address is 500 and w = 4.

Ans: *Length of three dimensions of A –*
L_1 = 12 – 4 + 1 = 9
L_2 = 1 – (-2) + 1 = 4
L_3 = 14 – 8 = 6

Therefore, A contains 9 X 4 X 6 = 216 elements

Now, E_1 = 5 – 4 = 1
     E_2 = 4 - (-2) = 8
     E_3 = 9 – 8 = 1

Row Major Order
     Address (5, 4, 9 = 500 + 4 (6 (4(1) + 8) + 1)
          = 500 + 4 (6 (12) + 1)
          = 500 + 4 (73)
     Address (5, 4, 9) = 500 + 292 = 792

Column Major Order
     Address (5, 4, 9) = 500 + 4 ((1.4 + 8) 9 + 1)
          = 500 + ((12)9 + 1)
     Address (5, 4, 9) = 500 + 145 = 645

// Write a program to read and display a 2 X 2 X 2 array.

```
import java.util.Scanner;
public class ThreeDimensionalArrayDemo {
```

```java
public static void main(String[] args) {

    Scanner scn = new Scanner(System.in);
    int[][][] array = new int[2][2][2];

    System.out.println("Enter the elements of array:");
    for (int i = 0; i < 2; i++) {
        for (int j = 0; j < 2; j++) {
            for (int k = 0; k < 2; k++) {
                array[i][j][k] = scn.nextInt();
            }
        }
    }

    System.out.println("The array is:");
    for (int i = 0; i < 2; i++) {
        System.out.println();
        for (int j = 0; j < 2; j++) {
            System.out.println();
            for (int k = 0; k < 2; k++) {
                System.out.println("array[" + i + "]" +
                        "[" + j + "]" + "[" + k + "]" +
                        " = " + array[i][j][k]);
            }
        }
    }
}
}
```

The output of the program is shown as:

3.12   Arrays and Their Applications

Arrays are very frequently used in Java, as they have various applications which are very useful. These applications include the following:

- Arrays are used for sorting the elements in ascending or descending order.

- Arrays are also used to implement various other data structures like stacks, queues, hash tables, and so on.

- Arrays are widely used to implement matrices, vectors, and various other kinds of rectangular tables.

- Various other operations can be performed on the arrays, which include searching, merging, sorting, and so forth.

Frequently Asked Questions
5. List some of the applications of arrays. **Ans:** *1. Arrays are very useful in storing the data in contiguous memory locations.* *2. Arrays are used for implementing various other data structures such as stacks, queues, and so on.* *3. Arrays are very useful, as we can perform various operations on them.*

3.13   Sparse Matrices

A sparse matrix is a matrix with a relatively high proportion of zero entries in it. A sparse matrix utilizes the memory space efficiently. Storing of null elements in the matrix is a waste of memory, so we adopt a technique to store only not-null elements in the sparse matrices.

For Example:

$$\begin{bmatrix} 0 & 0 & 6 & 0 & 0 \\ 1 & 0 & 0 & 0 & 0 \\ 0 & 0 & 0 & 0 & 2 \\ 0 & 5 & 0 & 0 & 0 \end{bmatrix}$$

Figure 3.7. Representation of a sparse matrix.

3.14   Types of Sparse Matrices

There are three types of sparse matrices, which are:

1. **Lower-triangular matrix:** In this type of sparse matrix, all the elements above the main diagonal must have a zero value, or in other words we can say that all the elements below the main diagonal should contain non-zero elements only. This type of matrix is called a lower-triangular matrix.

$$\begin{bmatrix} 5 & 0 & 0 & 0 \\ 4 & 6 & 0 & 0 \\ -3 & 9 & -5 & 0 \\ 2 & 1 & 7 & 3 \end{bmatrix}$$

Figure 3.8. Lower-triangular matrix.

2. **Upper-triangular matrix:** In this type of sparse matrix, all the elements above the main diagonal should contain non-zero elements only, or in other words we can say that all the elements below the main diagonal should have a zero value. This type of matrix is called an upper-triangular matrix.

$$\begin{bmatrix} 1 & 2 & 3 & 4 \\ 0 & 6 & -1 & 5 \\ 0 & 0 & -7 & 8 \\ 0 & 0 & 0 & 9 \end{bmatrix}$$

Figure 3.9. Upper-triangular matrix.

3. **Tri-diagonal matrix:** In this type, elements with a non-zero value can appear only on the diagonal or adjacent to the diagonal. This type of matrix is a tri-diagonal matrix.

$$\begin{bmatrix} 6 & 2 & 0 & 0 \\ 8 & 9 & -2 & 0 \\ 0 & 5 & -7 & 3 \\ 0 & 0 & 1 & 4 \end{bmatrix}$$

Figure 3.10. Tri-diagonal matrix.

3.15   Representation of Sparse Matrices

There are two ways in which sparse matrices can be represented, which are:

1. **Array Representation/3-Tuple Representation:** This representation contains three rows in which the first row represents the number of rows, columns, and non-zero entries/values in the sparse matrix. Elements in the other rows give information about the location and value of non-zero elements.

   For example, let us consider a sparse matrix.

$$\begin{bmatrix} 0 & 0 & 0 & 0 & 1 \\ 0 & 0 & 0 & 0 & 0 \\ 0 & 0 & 3 & 0 & 0 \\ 0 & 5 & 0 & 0 & 0 \end{bmatrix}$$

   An array representation of the previous sparse matrix will be:

Row	Column	Non-Zero Value
0	4	1
2	2	3
3	1	5

2. **Linked Representation:** A sparse matrix can also be represented in a linked way. In this representation we store the number of rows, columns, and non-zero entries in a single node, and there is an address field that stores the next location. Let us consider an example to understand more clearly.

   Let us consider a sparse matrix:

$$\begin{bmatrix} 0 & 6 & 0 & 0 & 0 \\ 0 & 0 & 0 & 0 & 0 \\ 0 & 0 & 0 & 8 & 3 \end{bmatrix}$$

   Linked representation of the previous sparse matrix will be as follows:

| 3 | 5 | 3 | |

No. of rows

No. of columns

Non zero values

Address of next loaction

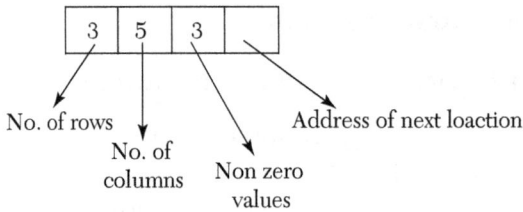

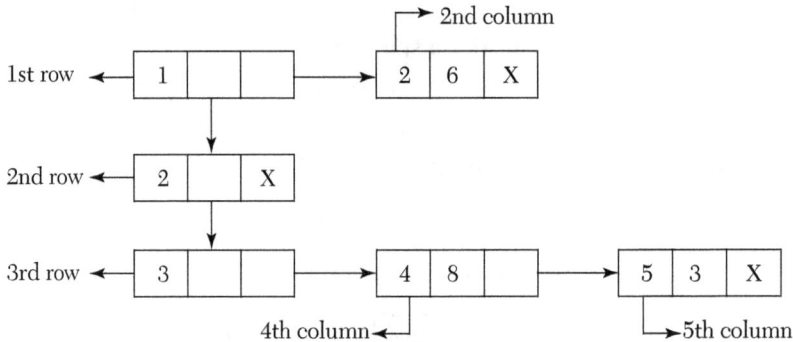

Figure 3.12. Linked representation of a sparse matrix.

Frequently Asked Questions

6. Explain the sparse matrix.

Ans: *A matrix in which the number of zero entries is much higher than the number of non-zero entries is called a sparse matrix. The natural method of representing matrices in memory as two-dimensional arrays may not be suitable for sparse matrices. One may save space by storing only non-zero entries. We can represent a sparse matrix by using a three-tuple method of storage:*

- *Row Major Method*
- *Column Major Method*

3.16   Summary

- An array is a collection of homogeneous (similar) types of data elements in contiguous memory. An array is a linear data structure because all elements of an array are stored in linear order.

- An array must be declared before it is used.

- The initialization of the elements of an array at compile time is done in the same way as when we initialize the normal or ordinary variables at the time of their declaration.

- Initialization of elements of an array at runtime refers to the method of inputting the values from the keyboard.

- The address of the elements in a 1-D array can be calculated very easily, as an array stores all its data elements in contiguous memory locations, storing the base address.

- Traversing an array means to access each and every element in an array exactly once so that it can be processed.

- Insertion of an element in an array refers to the operation of adding an element to the array. It can be done in three ways.

- Deleting an element from an array refers to the operation of the removal of an element from an array. Deletion is also done in three ways.

- Searching for an element in an array means finding whether a particular value exists in an array or not. If that particular value is found, then the searching is said to be successful and the position/location of that particular value is returned. If the value is not found, then searching will be said to be unsuccessful.

- A linear search is a very simple technique used to search for a particular value in an array.

- The merging of two arrays means copying the elements of the first and second arrays into a third array.

- Sorting an array means arranging the data elements of a data structure in a specified order either in ascending or descending order.

- Selection sort is a sorting technique that works by finding the smallest value in the array and placing it in the first position. After that, it then finds the second smallest value and places it in the second position. This process is repeated until the whole array is sorted.

- Unlike one-dimensional arrays, 2-D arrays are organized in the form of grids or tables. They are collections of 1-D arrays.

- For declaring two-dimensional arrays, we must know the name of the array, the data type of each element, and the size of each dimension (size of row and column).

- A multidimensional array is also known as an n-dimensional array. It is an array of arrays. It has n indices in it, which also justifies its name of an n-dimensional array.

- A sparse matrix is a matrix with a relatively high proportion of zero entries in it. A sparse matrix is used because it utilizes the memory space efficiently.

3.17 Exercises

3.17.1 Theory Questions

1. How do you define an array and how is it represented in the memory?

2. What are the various operations that can be performed on arrays? Discuss in detail.

3. Explain the concept of two-dimensional arrays.

4. In how many ways can arrays be initialized? Explain in detail.

5. Consider a one-dimensional array declared as int[] arr = new int[10], and calculate the address of arr [7] if the base address is 200 and the size of each element is 2.

6. What do you mean by sorting an array? Explain.

7. Write an algorithm to perform the selection sort technique. Give the advantages of using the selection sort technique for sorting the elements in an array.

8. Explain the process of merging two arrays along with the algorithm.

9. Consider a two-dimensional array declared as int[][] array = new int[10][10], and calculate the address of the element array [5] [6] if the base address = 10000 and the size of each element = 2, assuming the elements are to be stored in column major order.

10. Give some of the applications of arrays.

11. What do you understand by a linear search? Give the algorithm.

12. What is a sparse matrix? Also explain its types.

13. Consider a three-dimensional array A (2:6, -1:7, 9:10) and calculate the address of A (9, 6, 8) using row major order and column major order, where the base address is 2000 and w = 4.

14. Explain the linked representation of sparse matrices in detail.

15. Write the formulae for calculating the addresses of elements in row major and column major order in 2-D and 3-D arrays.

3.17.2 Programming Questions

1. Write a Java program to traverse an entire array.

2. Write a Java program to perform insertion at a specified position in a one-dimensional array.

3. Write a Java program to multiply two matrices.

4. Write a Java program which reads a matrix and displays the
- **(a)** Sum of its rows' elements
- **(b)** Sum of its columns' elements
- **(c)** Sum of its diagonal's elements

5. Write a Java program to perform the deletion of an element from the beginning.

6. Write a menu-driven Java program to perform various insertions and deletions in an array using the switch case.

7. Write a program to read and display a square matrix using class.

8. Write an algorithm for reversing the array.

9. Write a program which reads an array of fifty integers. Display all the pairs of elements whose sum is 25.

10. Write a Java program to read an array of ten integers and then find the smallest and largest numbers in the array.

11. Write a Java program to add two sparse matrices using classes.

3.17.3 Multiple Choice Questions

1. If an array is declared as int[][] array = new int[20] [20], then how many elements can it store?
- **(a)** 20
- **(b)** 40
- **(c)** 400
- **(d)** None of these

2. The elements of an array are always stored in _____ memory locations.

   (a) Random

   (b) Sequential

   (c) Both

   (d) None of these

3. What will be the output of the given program?

```java
public class Demo {
    public static void main(String[] args) {
        int[] arr = new int[5];
        int a;
        int b;
        int c;
        for (int i = 1; i <= 5; i++)
            arr[i - 1] = 5 * i;
        a = ++arr[1];
        b = a++;
        c = arr[2]++;
        System.out.println(a + ", " + b + ", " + c);
    }
}
```

   (a) 11, 12, 15

   (b) 12, 11, 15

   (c) 13, 12, 15

   (d) 12, 13, 20

4. What will be the output of the following program after execution?

```java
public class Demo {
    public static void main(String[] args) {
        int[] arr = new int[5];
        for (int i = 1; i <= 5; i++)
            arr[i - 1] = 5 * i;
        System.out.println(arr[6]);
    }
}
```

(a) 10

(b) Garbage value

(c) Runtime error

(d) None of these

5. If an array is declared as int[] array = new int [], then the nth element can be accessed by:

(a) array[n]

(b) *(array + n)

(c) *(n + array)

(d) None of these

(e) All of these

6. Array [5] = 19 initializes the _____ element of the array with value 19.

(a) 4th

(b) 5th

(c) 6th

(d) 7th

7. By default, the first subscript of the array is _____.

(a) 2

(b) 1

(c) -1

(d) 0

8. A multidimensional array, in simple terms, is an

(a) array of arrays

(b) array of addresses

(c) Both

(d) None of the above

9. A loop is used to access all the elements of an array.

(a) False

(b) True

(c) None of the above

10. Declaring an array means specifying the _____, _____, and _____.

    (a) Data type, name, size

    (b) Data elements, name, data type

    (c) Name, size, address

    (d) All of the above

11. A sparse matrix has a _____.

    (a) High proportion of zeroes

    (b) Low proportion of zeroes

    (c) Both (a) and (b)

    (d) None of the above

CHAPTER 4

LINKED LISTS

4.1 Introduction

We have already learned that an array is a collection of data elements stored in contiguous memory locations. Also, we studied that arrays were static in nature; that is, the size of the array must be specified when declaring an array, which limits the number of elements to be stored in the array. For example, if we have an array declared as int[] array = new int[15], then the array can contain a maximum of fifteen elements and not more than that. This method of allocating memory is good when the exact number of elements is known, but if we are not sure of the number of elements, then there will be a problem, as in data structures our aim is to make programs efficient by consuming less memory space along with minimal time. To overcome this problem, we will use linked lists.

4.2 Definition of a Linked List

A linked list is a linear collection of data elements. These data elements are called nodes, and they point to the next node. A linked list is a data structure which can be used to implement other data structures such as stacks, queues, trees, and so on. *A linked list is a sequence of nodes in which each node contains one or more than one data field which points to the next node.* Also, linked lists are dynamic in nature; that is, memory is allocated as and when required. There is no need to know the exact size or exact number of elements as in the case of arrays. The following is an example of a simple linked list which contains five nodes:

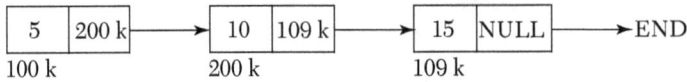

Figure 4.1. A linked list.

In the previous figure, we have made a linked list in which each node is divided into two parts:

1. *The first part contains the information/data.*

2. *The second part contains the address of the next node.*

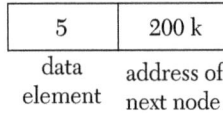

The last node will not have any next node connected to it, so it will store a special value called NULL. Usually NULL is defined by -1. Therefore, the NULL node represents the end of the linked list. Also, there is another special node START that stores the address of the first node of the linked list. Therefore, the START node represents the beginning of the linked list. If START = NULL, then it means that the linked list is empty. A linked list, since each node points to another node which is of the same type, is known as a self-referential data type or a self-referential structure.

The self-referential structure in a linked list is as follows:

```
private class Node {
      int info;
      Node next;
}
```

Practical Application

- A simple real-life example is how each coach on a train is connected to its previous and next coach (except the first and last). In terms of programming, consider the coach body as a node and the connectors as links to the previous and next nodes.

- The brain is also a good example of a linked list. In the initial stages of learning something by heart, the natural process is to link one item to another item. It's a subconscious act. Also, when we forget something and try to remember, then our brain follows associations and tries to link one memory with another and so on until we finally recall the lost memory.

Frequently Asked Questions

1. Define linked list.

Ans: *A linked list is a linear collection of data elements, called nodes, where the linear order is given by means of nodes. It is a dynamic data structure. For every data item in a linked list, there is an associated node that gives the memory location of the next data item in the linked list. The data items in the linked list are not in consecutive memory locations.*

2. List the advantages and disadvantages of a linked list.

Ans:

Advantages of linked lists

1. *Linked lists are dynamic data structures; that is, they can grow or shrink during the execution of the program.*

2. *Linked lists have efficient memory utilization. Memory is allocated whenever it is required, and it is de-allocated whenever it is no longer needed.*

3. *Insertion and deletion are easier and efficient.*

4. *Many complex applications can be easily carried out with linked lists.*

Disadvantages of linked lists

1. *They consume more space because every node requires an additional node to store the address of the next node.*

2. *Searching a particular element in the list is difficult and time-consuming.*

4.3 Memory Allocation in a Linked List

The process or concept of linked lists supports dynamic memory allocation. Now, what is meant by dynamic memory allocation? The answer to this simple question is that the process of allocating memory during the execution of the program or the process of allocating memory to the variables at runtime is called dynamic memory allocation. Until now we have studied arrays in which we declared the size of the array initially, such as array[50]. This statement after execution allocates the memory for fifty integers. But there can be a problem if we use only 30% of the memory and the rest of the allocated memory is wasted. Therefore, to overcome this problem of wastage of memory space or, in other words, to utilize the memory efficiently, dynamic memory allocation is used, which allows us to allocate/reserve the memory that is actually required. Hence, it will overcome the problem of wastage of memory space as in the case of arrays. Dynamic memory allocation is best when we are not aware of the memory requirements in advance.

4.4 Types of Linked Lists

There are different types of linked lists that will be discussed in this section. These include:

1. Singly Linked List

2. Circular Linked List

3. Doubly Linked List

4. Header Linked List

Now we will discuss all of them in detail.

4.4.1 Singly Linked List

A singly linked list is the simplest type of linked list, in which each node contains some information/data and only one node which points to the next node in the linked list. The traversal of data elements in a singly linked list can be done only in one way.

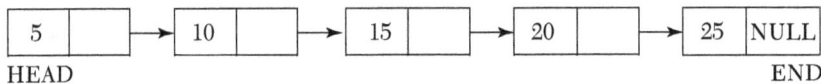

HEAD   END

Figure 4.2. Singly linked list.

4.4.2 Operations on a Singly Linked List

Various operations can be performed on a singly linked list, which include:

- Traversing a linked list
- Searching for a given value in a linked list
- Inserting a new node in a linked list
- Deleting a node from a linked list
- Concatenation of two linked lists
- Sorting a linked list
- Reversing a linked list

Let us now discuss all these operations in detail.

(a) Traversing a linked list

Traversing a linked list means accessing all the nodes of the linked list exactly once. A linked list will always contain a START node, which stores the address of the first node of the linked list and which also represents the beginning of the linked list, and a NULL node, which represents the end of the linked list. For traversing a linked list, we will use another node variable "node" which will point to the node that is currently being accessed. The algorithm for traversing a linked list is shown as follows:

Algorithm for traversing a linked list

```
Step 1: Set NODE = START
Step 2: Repeat Steps 3 & 4 while NODE != NULL
Step 3: Print NODE. INFO
Step 4: Set NODE = NODE. NEXT
        [End of Loop]

Step 5: Exit
```

(b) Searching for a given value in a linked list

Searching for a value in a linked list means to find a particular element/value in the linked list. As we discussed earlier, a node in a linked list contains two parts; one part is the information part and the other is the address part. Hence, searching refers to the process of finding whether or not the given value exists in the information part of any node. If the value is present, then the address of that particular value is returned and the search is said to be successful; otherwise, the search is unsuccessful. A linked list will always contain a START node, which stores the address of the first node of the linked list and also represents the beginning of the linked list, and a NULL node, which represents the end of the linked list. There is another variable NODE which will point to the current node being accessed. SEARCH_ VAL is the value to be searched in the linked list, and POS is the position/ address of the node at which the value is found. The algorithm for searching a value in a linked list is given as follows:

Algorithm to search a value in a linked list

```
Step 1: Set NODE = START
Step 2: Repeat Step 3 while NODE != NULL
Step 3: IF SEARCH_VAL = NODE. INFO
                Set POS = NODE
```

```
            Print Successful Search!!
            Go to Step 5
    [End of If]
    ELSE
            Set NODE = NODE. NEXT
    [End of Loop]
Step 4: Print Unsuccessful Search!!
Step 5: Exit
```

For example, if we have a linked list and we are searching for 15 in the list, then the steps are shown as follows:

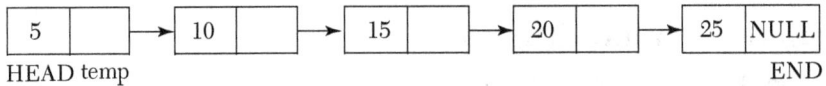

The value to be searched is 15, but temp.data = 5. Hence we will move forward towards the next node temp.next

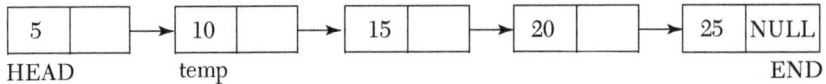

The value to be searched is 15, but temp.data = 10. Hence we will move forward towards the next node temp.next

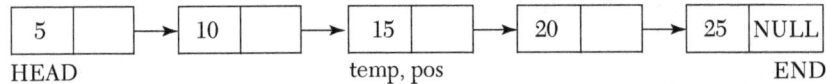

The value to be searched is 15, but temp.data = 15.
Hence search is successful return the pos.

Figure 4.3. An example of searching a linked list.

(c) Inserting a new node in a linked list

Here, we will learn how a new node is inserted in an existing linked list. We will discuss three cases in the insertion process, which include:

1. A new node is inserted at the beginning of the linked list.

2. A new node is inserted at the end of the linked list.

3. A new node is inserted after the given node in a linked list.

Let us now discuss all of these cases in detail.

1. Inserting a new node in the beginning of a linked list

In the case of inserting a new node in the beginning of a linked list, we will first check the overflow condition, which is whether the memory is available for a new node. If the memory is not available, then an overflow message is displayed; otherwise, the memory is allocated for the new node. Now, we will initialize the node with its info part, and its address part will contain the address of the first node of the list, which is the START node. Hence, the new node is added as the first node in the list, and the START node will point to the first node of the list. Now to understand better, let us take an example. Consider a linked list as shown in the following figure with five nodes; a new node will be inserted in the beginning of the linked list.

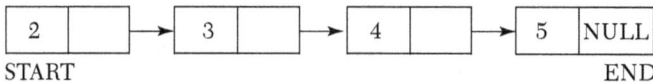

Now we need to insert 1 at the start of the linked list. First we create a newNode with newNode data=1 and newNode=NULL

Now as we need to insert the newNode in the beginning of the linked list so we assign newNode next=START. This way the next of newNode will contain the memorty address of START.

Now assign START as newNode so that whenever we access START it shows 1.

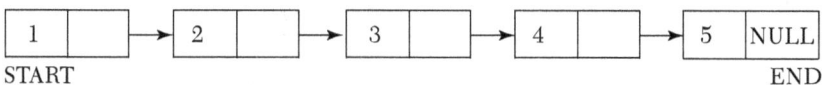

Figure 4.4. Inserting a new node at the beginning of a linked list.

From the previous example, it is clear how a new node will be inserted in an already existing linked list. Let us now understand its algorithm:

Algorithm for inserting a new node in the beginning of a linked list

```
Step 1: START
Step 2: IF NODE = NULL
            Print OVERFLOW
            Go to Step 8
        [End of If]
```

```
Step 3:  Set NEW NODE = NODE
Step 4:  Set NODE = NODE. NEXT
Step 5:  Set NEW NODE. INFO = VALUE
Step 6:  Set NEW NODE. NEXT = START
Step 7:  Set START = NEW NODE
Step 8:  EXIT
```

2. Inserting a new node at the end of a linked list

To insert the new node at the end of the linked list, we will first check the overflow condition, which is whether the memory is available for a new node. If the memory is not available, then an overflow message is displayed; otherwise, the memory is allocated for the new node. Then a NODE variable is made which will initially point to START and will be used to traverse the linked list until it reaches the last node. When it reaches the last node, the NEXT part of the last node will store the address of the new node, and the NEXT part of the NEW NODE will contain NULL, which will denote the end of the linked list. Let us understand this with the help of an algorithm:

Algorithm for inserting a new node at the end of a linked list

```
Step 1:  START
Step 2:  IF NEW NODE = NULL
             Print OVERFLOW
         [End of If]
Step 3:  Set NEW NODE. INFO = VALUE
Step 4:  Set NEW NODE. NEXT = NULL
Step 5:  Set TEMP = START
Step 6:  Repeat Step 6 while TEMP. NEXT != NULL
             Set TEMP = TEMP. NEXT
         [End of Loop]
Step 7:  Set TEMP. NEXT = NEW NODE
Step 8:  EXIT
```

From the previous algorithm we understand how to insert a new node at the end of already existing linked list. Now we will study further with the help of an example. Consider a linked list as shown in the following figure with four nodes; a new node will be inserted at the end of the linked list:

```
1 |   | --> 2 |   | --> 3 |   | --> 4 | NULL
```
START temp   END

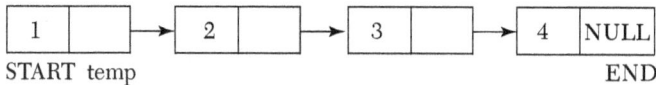

Now we have to add 5 at the end of the linked list. First we create newNode with newNode data=5 and newNode next=NULL.

```
5 | NULL | newNode
```

Create a node temp=START that will traverse the linked list till temp.next=NULL. Once it reached the last node or END, we can assign temp.next=newNode. This way we can insert a node at the end of a linked list.

START   temp   END
```
1 |   | --> 2 |   | --> 3 |   | --> 4 |   | --> 5 | NULL
```

Or we can simply keep a END node that refers to the last node of the linked list and assign ENDnext=newNode.

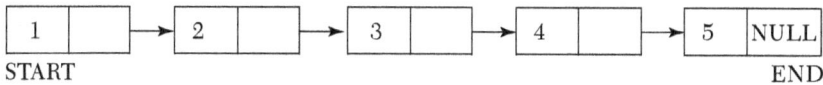

```
1 |   | --> 2 |   | --> 3 |   | --> 4 |   | --> 5 | NULL
```
START   END

Figure 4.5. Inserting a new node at the end of a linked list.

3. Inserting a new node after a node in a linked list

In this case, a new node is inserted after a given node in a linked list. As in the other cases, we will again check the overflow condition. If the memory for the new node is available, it will be allocated; otherwise, an overflow message is printed. Then a NODE variable is made which will initially point to START, and the NODE variable is used to traverse the linked list until it reaches the value/node after which the new node is to be inserted. When it reaches that node/value, then the NEXT part of that node will store the address of the new node and the NEXT part of the NEW NODE will store the address of its next node in the linked list. Let us understand this with the help of an example. Consider a linked list with four nodes, and a new node is to be inserted after the given node:

START

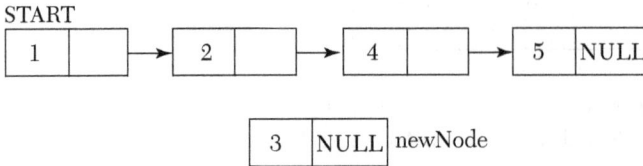

Now we have to insert 3 after 2 in the linked list. So we create a newNode with newNode data=3 and newNode next=NULL and a node prev=START

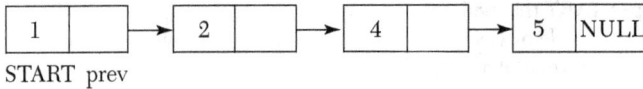

START prev

Now we will traverse on the linked list to check if prev.data is equal to 2 or not.

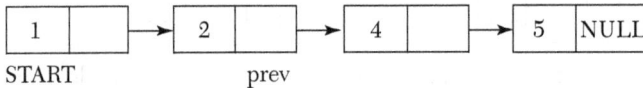

START   prev

So when prev.data=2 then we assign newNode.next=prev.next and next and prev.next-=newNode. This why we insert a node after a node in linked list.

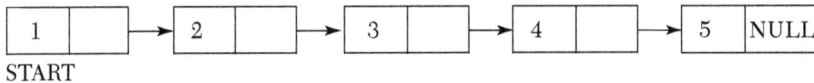

START

Figure 4.6. Inserting a new node after a given node in a linked list.

From the previous example, we learned how a node can be inserted after a given node. Now we will understand this with the help of an algorithm.

Algorithm for inserting a new node after a given node in a linked list

```
Step 1: START
Step 2: IF NEW NODE = NULL
            Print OVERFLOW
        [End of If]
Step 3: Set NEW NODE. INFO = VALUE
Step 4: Set NEW NODE. NEXT = NULL
Step 5: Set PREV = NODE
Step 6: Repeat Step 6 while PREV. INFO != GIVEN_VAL
            Set PREV = PREV. NEXT
        [End of Loop]
Step 9: Set NEW NODE. NEXT = PREV
Step 10: Set PREV. NEXT = NEW NODE

Step 11: EXIT
```

(d) Deleting a node from a linked list

In this section, we will learn how a node is deleted from an already existing linked list. We will discuss three cases in the deletion process, which include the following:

1. A node is deleted from the beginning of the linked list.

2. A node is deleted from the end of the linked list.

3. A node is deleted after a given node from the linked list.

Let us now discuss all of these cases in detail.

1. Deleting a node from the beginning of the linked list

In the case of deleting a node from the beginning of a linked list, we will first check the underflow condition, which occurs when we try to delete a node from a linked list which is empty. This situation exists when the START node is equal to NULL. If the condition is true, then the underflow message is printed on the screen; otherwise, the node is deleted from the linked list. Consider a linked list as shown in the following figure with five nodes; the node will be deleted from the beginning of the linked list.

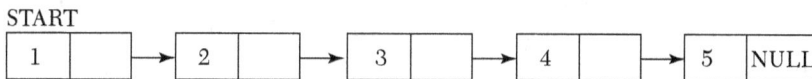

START
| 1 | | → | 2 | | → | 3 | | → | 4 | | → | 5 | NULL |

Now we need to delete 1 from the linked list. That means we need to delete the START node.

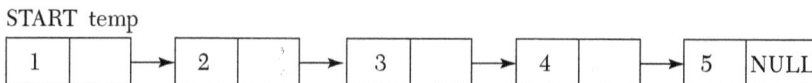

START temp
| 1 | | → | 2 | | → | 3 | | → | 4 | | → | 5 | NULL |

So first we will declare a node temp and assign it equals to START, i.e., temp=START. Then we will assign START=temp.next. Then temp.next=NULL. This will remove the first node of the linked list.

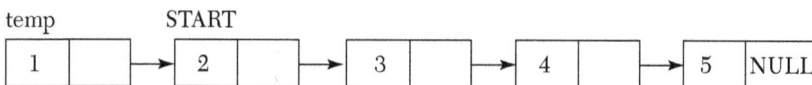

temp   START
| 1 | | → | 2 | | → | 3 | | → | 4 | | → | 5 | NULL |

We can also simply assign START=START.next. The frist node will automatically become unreachable as the first node will be START itself.

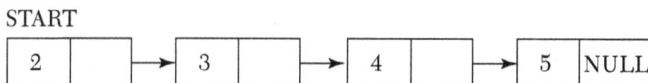

START
| 2 | | → | 3 | | → | 4 | | → | 5 | NULL |

Figure 4.7. Deleting a node from the beginning of a linked list.

From the previous example, it is clear how a node is deleted from an already existing linked list. Let us now understand its algorithm:

Algorithm for deleting a node from the beginning of a linked list

```
Step 1: START
Step 2: IF START = NULL
            Print UNDERFLOW
        [End Of If]
```

Step 3: Set START = START. NEXT

Step 4: EXIT

In the previous algorithm, first we check for the underflow condition, that is, whether there are any nodes present in the linked list or not. If there are no nodes, then an underflow message will be printed; otherwise, we move to Step 3 where we are initializing NODE to START; that is, NODE will now store the address of the first node. In the next step START is moved to the second node, as now START will store the address of the second node. Hence, the first node is deleted and the memory which was occupied by NODE (initially the first node of the list) is free.

2. Deleting a node from the end of the linked list

In the case of deleting a node from the end of the linked list, we will first check the underflow condition. This situation exists when the START node is equal to NULL. Hence, if the condition is true, then the underflow message is printed on the screen; otherwise, the node is deleted from the linked list. Consider a linked list as shown in the following figure with five nodes; the node will be deleted from the end of the linked list.

START

| 1 | | → | 2 | | → | 3 | | → | 4 | | → | 5 | NULL |

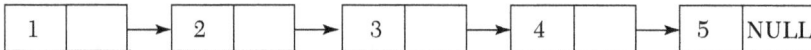

So we need to delete 5 from the linked list. For that the last mode shuld be deleted.

START temp

| 1 | | → | 2 | | → | 3 | | → | 4 | | → | 5 | NULL |

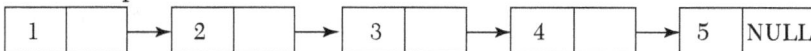

First declare a node temp=START and a node prev. This temp will traverse the linked list till temp.next=null and update prev=temp. When temp.next=NULL.prev will be the node just before the temp.

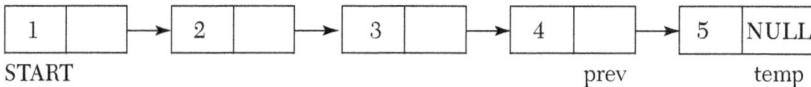

| 1 | | → | 2 | | → | 3 | | → | 4 | | → | 5 | NULL |

START   prev   temp

Assign prev.next=NULL. This way the last node will become unreachable and the last node will be removed from the linked list.

START

| 1 | | → | 2 | | → | 3 | | → | 4 | NULL |

Figure 4.8. Deleting a node from the end of a linked list.

Let us now understand the algorithm of deleting a node from the end of a linked list.

Algorithm for deleting a node from the end of a linked list

```
Step 1: START
Step 2: IF START = NULL
            Print UNDERFLOW
        [End Of If]
Step 3: Set PREV = START
Step 4: Repeat while TEMP. NEXT != NULL
            Set PREV = TEMP
            Set TEMP = TEMP. NEXT
        [End of Loop]
Step 5: Set PREV. NEXT = NULL
Step 6: EXIT
```

In the previous algorithm, we again check for the underflow condition. If the condition is true, then the underflow message is printed; otherwise, NODE is initialized to the START node; that is, NODE is pointing to the first node of the list. In the loop we have taken another node variable PREV, which will always point to one node before the NODE node. After reaching the last node of the list, we will set the next part of PREV to NULL. Therefore, the last node is deleted, and the memory which was occupied by the NODE node is now free.

3. Deleting a node after a given node from the linked list

In the case of deleting a node after a given node from the linked list, we will again check the underflow condition that we checked in both the other cases. This situation exists when the START node is equal to NULL. Hence, if the condition is true, then the underflow message is printed; otherwise, the node is deleted from the linked list. Consider a linked list as shown in the following figure with five nodes initially; the node will be deleted after a given node from the linked list.

START

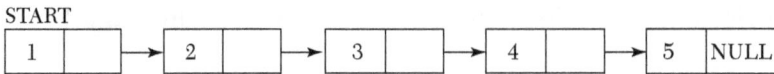

So we need to remove 3 from the linked list. For that we need to remove the node between two nodes.

START prev   temp

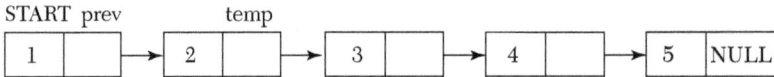

First we need to declare a node prev=START and temp=prev.next. This prev and temp will traverse the linked list till prev.data=2. Once we reach the node 2 we need establish a link between prev and temp.next. So we will assign prev.next=temp.next.

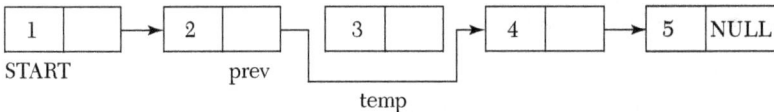

START   prev

temp

This way 3 will become unreachable and will be removed from the linked list.

START

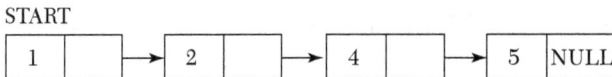

Figure 4.9. Deleting a node after a given node from the linked list.

Now let us understand the previous case with the help of an algorithm.

Algorithm for deleting a node after a given node from the linked list

```
Step 1: START
Step 2: IF START = NULL
            Print UNDERFLOW
       [End Of If]
Step 3: Set TEMP = START
Step 4: Set PREV = START
Step 5: Repeat while PREV. INFO != GIVEN_VAL
            Set PREV = TEMP
            Set TEMP = TEMP. NEXT
       [End of Loop]
Step 6: Set PREV. NEXT = TEMP. NEXT
Step 7: EXIT
```

In the previous algorithm, we are first checking for the underflow condition. If the condition is true, then the underflow message is printed. Otherwise, NODE is initialized to the START node; that is, NODE is pointing to the first node of the list. In the loop we have taken another node variable PREV, which will always point one node before the NODE node. After reaching the node containing the given value that is to be deleted, we will set the next node of the node containing the given value to the address contained in the next part of the succeeding node. Therefore, the node is deleted, and the memory that was being occupied by NODE is now free.

(e) Concatenation of two linked lists

A concatenated linked list is created by the process of concatenating two different-sized linked lists into one linked list. Let us understand the concept of concatenation with the help of a function:

```
public void concatenate(Node node1, Node node2) {
    Node temp = new Node();
    temp = node1;
    while (temp.next != null) {
        temp = temp.next;
    }
    temp.next = node2;
}
```

(f) Sorting a linked list

Sorting is the process of arranging the data elements in a sequence, either in ascending order or in descending order. In this we are arranging the information of the linked list in a sequence.

(g) Reversing a linked list

In the process of reversing a linear linked list, we will take three node variables, that is, PREV, NODE, and NEW, which will hold the addresses of the previous node, current node, and the next node respectively in the linked list. We will begin with the address of the first node that is held in another node variable START which is assigned to NODE, and PREV is assigned to NULL. Now, let us understand it with the help of a function:

```
public void reverse_list() {
    Node = start;
    Node prev = null;
    while (node != null) {
        Node newnode = node.next;
        node.next = prev;
        prev = node;
        node = newnode;
    }
}
```

//Write a menu-driven program for singly linked lists performing insertion and deletion of all cases.

```
public class LinkedList {
    private class Node {
        int data;
        Node next;
```

```
    }
    private Node start;
    private int size;
    private Node create_new_node(int item) throws Exception {
        Node = new Node();
        if (node == null) {
            throw new Exception("Memory not allocated");
        } else {
            node.data = item;
            node.next = null;
            return node;
        }
    }
    public void insertion(int item, int pos) throws Exception {
        if (pos < 0 || pos > size) {
            throw new Exception("Invalid Index");
        }
        if (pos == 0) {
            insertion_at_beginning(item);
        } else if (pos == size) {
            insertion_at_end(item);
        } else {
            Node = create_new_node(item);
            Node prev = getNodeAt(pos - 1);
            node.next = prev.next;
            prev.next = node;
            size++;
        }
    }
    private void insertion_at_beginning(int item) throws Exception {
        Node = create_new_node(item);
        Node temp = new Node();
        if (start == null) {
            start = node;
            start.next = null;
        } else {
            temp = start;
            start = node;
            start.next = temp;
        }
        size++;
    }
    private void insertion_at_end(int item) throws Exception {
        Node = create_new_node(item);
```

```
        Node temp = new Node();
        temp = start;
        while (temp.next != null) {
            temp = temp.next;
        }
        temp.next = node;
        size++;
    }
    public void deletion(int pos) throws Exception {
        if (pos < 0 || pos == size) {
            throw new Exception("Invalid Index");
        }
        if (size == 0) {
            throw new Exception("Linked List is Empty");
        }
        if (pos == 0) {
            deletion_at_beginning();
        } else if (pos == size - 1) {
            deletion_at_end(pos);
        } else {
            Node prev = getNodeAt(pos - 1);
            Node next = getNodeAt(pos + 1);
            prev.next = next;
        }
        size--;
    }
    private void deletion_at_beginning() {
        start = start.next;
    }
    private void deletion_at_end(int pos) throws Exception {
        Node prev = getNodeAt(pos - 1);
        prev.next = null;
    }
    public void display() throws Exception {
        if (size == 0) {
            throw new Exception("Linked List is Empty");
        }
        Node temp = start;
        while (temp != null) {
            System.out.print(temp.data + "->");
            temp = temp.next;
        }
        System.out.print("NULL \n");
    }
```

```java
    public Node getNodeAt(int idx) throws Exception {
        if (size == 0) {
            throw new Exception("Linked List is Empty");
        }
        if (idx < 0 || idx >= size) {
            throw new Exception("Invalid Index");
        }
        Node temp = start;
        for (int i = 0; i < idx; i++) {
            temp = temp.next;
        }
        return temp;
    }
}


//CLIENT CLASS
import java.util.Scanner;
public class LLClient {
    public static void main(String[] args) {
        Scanner scn = new Scanner(System.in);
        LinkedList list = new LinkedList();
        boolean flag = true;
        try {
            while (flag) {
                System.out.println("\n***MENU***");
                System.out.println("1. Insertion");
                System.out.println("2. Deletion");
                System.out.println("3. Display");
                System.out.println("4. Exit");
                System.out.println("Enter your choice: ");
                int choice = scn.nextInt();
                int item = 0, pos = 0;
                switch (choice) {
                case 1:
                    System.out.println("Enter value of node: ");
                    item = scn.nextInt();
                    System.out.println("Enter position of
                                    node:");
                    pos = scn.nextInt();
                    list.insertion(item, pos - 1);
                    System.out.println(item +
                            " inserted successfully at " + pos);
                    break;
```

```
                case 2:
                    System.out.println("Enter position of
                                        node:");
                    pos = scn.nextInt();
                    list.deletion(pos - 1);
                    System.out.println("Item deleted
                                        successfully");
                    break;
                case 3:
                    list.display();
                    break;
                case 4:
                    flag = false;
    System.out.println("Terminated......");
                    break;

                default:
                    System.out.println("Wrong choice");
                    break;
                }
            }
        } catch (Exception e) {
            System.out.println(e.getMessage());

        } }
}
```

The output of the program is shown as:

```
2
23 inserted successfully at 2

***MENU***
1. Insertion
2. Deletion
3. Display
4. Exit
Enter your choice:
1
Enter value of node:
12
Enter position of node:
3
12 inserted successfully at 3

***MENU***
1. Insertion
2. Deletion
3. Display
4. Exit
Enter your choice:
1
Enter value of node:
34
```

```
Enter position of node:
2
34 inserted successfully at 2

***MENU***
1. Insertion
2. Deletion
3. Display
4. Exit
Enter your choice:
3
1->34->23->12->NULL

***MENU***
1. Insertion
2. Deletion
3. Display
4. Exit
Enter your choice:
2
Enter position of node:
1
Item deleted successfully
```

```
***MENU***
1. Insertion
2. Deletion
3. Display
4. Exit
Enter your choice:
3
34->23->12->NULL

***MENU***
1. Insertion
2. Deletion
3. Display
4. Exit
Enter your choice:
4
Terminated.....

C:\Program Files\Java\jdk-12.0.2\bin>
```

After discussing the singly linked list, we will now learn about another type of linked list, the circular linked list.

4.4.3 Circular Linked Lists

Circular linked lists are a type of singly linked list in which the address part of the last node will store the address of the first node, unlike in singly linked lists in which the address part of the last node stores a unique value, NULL. While traversing a circular linked list, we can begin from any node and traverse the list in any direction, because a circular linked list does not have a first or last node. The memory declarations for representing a circular linked list are the same as for a linear linked list.

START

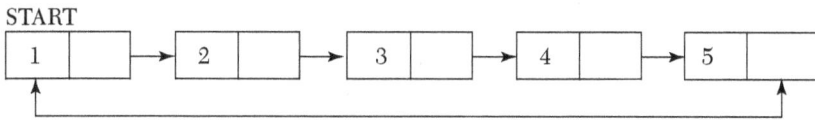

Figure 4.10. Circular linked list.

4.4.4 Operations on a Circular Linked List

Various operations can be performed on a circular linked list, which include:

(a) Inserting a new node in a circular linked list

(b) Deleting a node from a circular linked list

Let us now discuss both these cases in detail.

(a) Inserting a new node in a circular linked list

Here, we will learn how a new node is inserted in an existing linked list. We will discuss two cases in the insertion process, which include:

1. A new node is inserted at the beginning of the circular linked list.

2. A new node is inserted at the end of the circular linked list.

3. A new node is inserted after a given node (same as that for a singly linked list).

1. Inserting a new node in the beginning of a circular linked list

In the case of inserting a new node in the beginning of a circular linked list, we will first check the overflow condition, that is, whether the memory is available for a new node. If the memory is not available, then an overflow message is printed; otherwise, the memory is allocated for the new node. Now we will initialize the node with its info part, and its address part will contain the address of the first node of the list, which is the START node. Hence, the new node is added as the first node in the list, and the START node will point to the first node of the list. Now let us take an example. Consider a linked list as shown in the following figure with four nodes; a new node is to be inserted in the beginning of the circular linked list.

START temp

| 10 | | → | 15 | | → | 20 | | → | 25 | |

| 5 | NULL | newNode

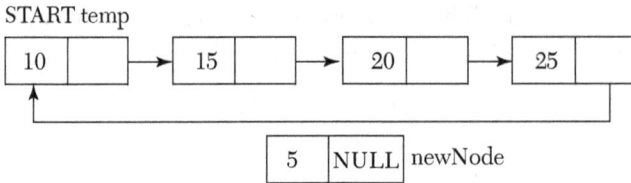

Here 5 is to be inserted at the beginning of the Circular Linked List.
First create a newNode with newNode.data=5 and newNode.next=NULL.
After that declare a node temp=START that will traverse the Circular
Linked List temp next=NULL.

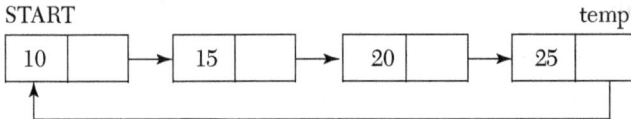

START   temp

| 10 | | → | 15 | | → | 20 | | → | 25 | |

When temp.next=START, assign newNode.next=START and
temp.next=newNode. Then newNode=START. This way we can insert a
node at the beginning of a Circular Linked List.

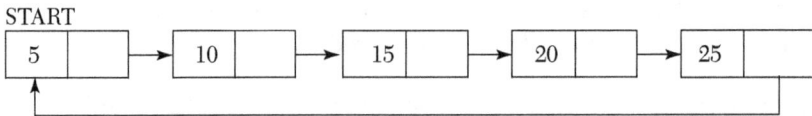

START

| 5 | | → | 10 | | → | 15 | | → | 20 | | → | 25 | |

Figure 4.11. Inserting a new node in the beginning of a circular linked list.

Now let us understand the previous case with the help of an algorithm.

Algorithm for inserting a new node in the beginning of a circular linked list

```
Step 1: START
Step 2: IF NEW NODE = NULL
            Print OVERFLOW
        [End Of If]
Step 3: Repeat while TEMP. NEXT != START
            Set TEMP = TEMP. NEXT
        [End of Loop]
Step 4: Set NEW NODE. INFO = VAL
Step 5: Set NEW NODE. NEXT = START
Step 6: Set TEMP. NEXT = NEW NODE
Step 7: Set START = NEW NODE
Step 8: EXIT
```

2. Inserting a new node at the end of a circular linked list

In this case, we will first check the overflow condition, that is, whether
the memory is available for a new node. If the memory is not available,
then an overflow message is printed; otherwise, the memory is allocated
for the new node. Then a NODE variable is made which will initially point

to START, and the NODE variable will be used to traverse the linked list until it reaches the last node. When it reaches the last node, the NEXT part of the last node will store the address of the new node, and the NEXT part of the NEW NODE will contain the address of the first node of the linked list, which is denoted by START. Let us understand it with the help of an algorithm:

Algorithm for inserting a new node at the end of a circular linked list

```
Step 1: START
Step 2: IF NEW NODE = NULL
            Print OVERFLOW
        [End Of If]
Step 3: Repeat while TEMP. NEXT != START
            Set TEMP = TEMP. NEXT
        [End of Loop]
Step 4: Set NEW NODE. INFO = VAL
Step 5: Set NEW NODE. NEXT = START
Step 6: Set TEMP. NEXT = NEW NODE
Step 7: EXIT
```

Let us take an example to understand it. Consider a linked list as shown in the following figure with four nodes; a new node is to be inserted at the end of the circular linked list.

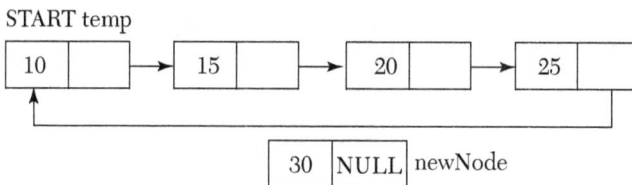

To insert 30 at the end of the Circular Linked List. First create a newNode with newNode.data=30 and newNode.next=NULL. Declare a node temp=START that will traverse the Circular Linked List till temp.next=START.

When temp.next is equal to START assign newNode.next=START and temp.next=newNode.

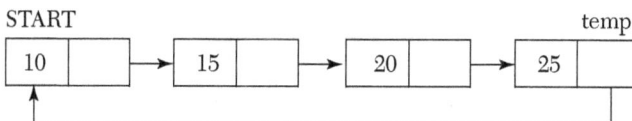

This way we can insert a node at the end of Circular Linked List.

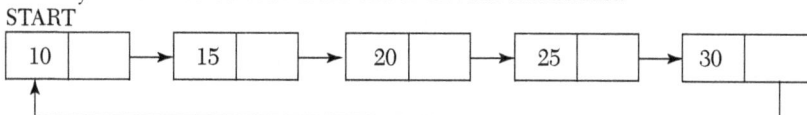

Figure 4.12. Inserting a new node at the end of a circular linked list.

(b) Deleting a node from a circular linked list

In this section, we will learn how a node is deleted from an already existing circular linked list. We will discuss two cases in the deletion process which include the following:

1. A node is deleted from the beginning of the circular linked list.

2. A node is deleted from the end of the circular linked list.

3. A node is deleted after a given node (same as that for a singly linked list).

1. Deleting a node from the beginning of a circular linked list

In the case of deleting a node from the beginning of a linked list, we will first check the underflow condition, which occurs when we try to delete a node from a linked list that is empty. This situation exists when the START node is equal to NULL. Hence, if the condition is true, then an underflow message is displayed; otherwise, the node is deleted from the linked list. Consider a linked list as shown in the following figure with four nodes; the first node will be deleted from the linked list.

START temp

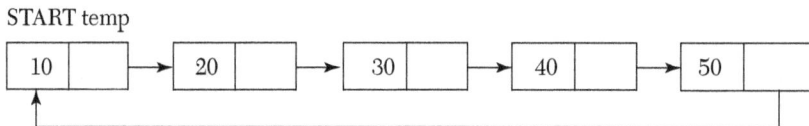

To remove 10 from the Circular Linked List, first declare a node temp=START. This temp will traverse the Circular Linked List till temp.next=START.

START   temp

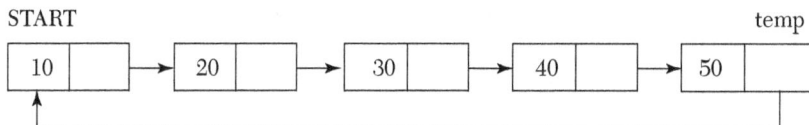

When temp.next=START, assign START=START.next and temp.next=START. Or we can just maintain an END node and assign END.next=START.next and START=START.next. This way we can delete a node from the end of a Circular Linked List.

START

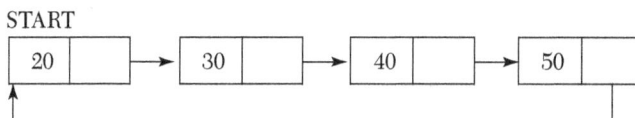

Figure 4.13. Deleting a node from the beginning of a circular linked list.

From the previous example, it is clear how a node will be deleted from an already existing linked list. Let us now understand its algorithm:

Algorithm for deleting a node from the beginning of a circular linked list

```
Step 1: START
Step 2: IF START = NULL
            Print UNDERFLOW
        [End Of If]
Step 3: Repeat while TEMP. NEXT != START
            Set TEMP = TEMP. NEXT
        [End of Loop]
Step 4: Set TEMP.NEXT = START.NEXT
Step 5: Set START = START.NEXT
Step 6: EXIT
```

The previous algorithm shows how a node is deleted from the beginning of the linked list. First, we check with the underflow condition. Now a node variable NODE is used, which will traverse the entire list until it reaches the last node of the list. Now, we change the next part of NODE to store the address of the second node of the list. Hence, the memory that occupied the first node is freed. Finally, the second node now becomes the first node of the linked list.

2. Deleting a node from the end of a circular linked list

In this case, we will first check the underflow condition, which is when we try to delete a node from a linked list that is empty. This situation occurs when the START node is equal to NULL. Hence, if the condition is true, then an underflow message is printed; otherwise, the node is deleted from the linked list. Consider a linked list as shown in the following figure with four nodes; the last node will be deleted from the linked list.

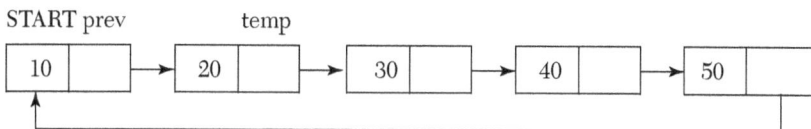

To delete 50 from the Circular Linked List, first declare a node prev=START and temp=prev.next. This temp will traverse the Circular Linked List till temp.next!=START.

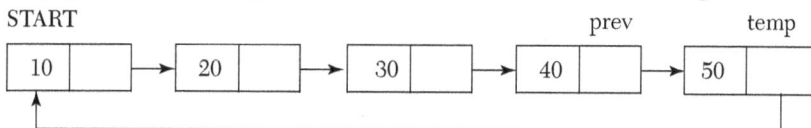

When temp.next=START, assign prev.next=temp.next. This way 50 will become unreachable and will be deleted from the Circular Linked List.

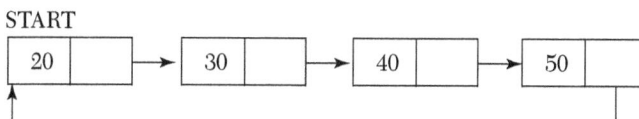

Figure 4.14. Deleting a node from the end of a circular linked list.

Let us now understand its algorithm:

Algorithm for deleting a node from the end of a circular linked list

```
Step 1: START
Step 2: IF START = NULL
            Print UNDERFLOW
        [End Of If]
Step 3: Set TEMP = START
Step 4: Repeat while TEMP. NEXT != START
            Set PREV = TEMP
            Set TEMP = TEMP. NEXT
        [End of Loop]
Step 5: Set PREV. NEXT = TEMP. NEXT
Step 6: EXIT
```

The previous algorithm shows how a node is deleted from the end of the linked list. First, we check with the underflow condition. Now a node variable NODE is used to traverse the entire list until it reaches the last node of the list. In the while loop we will use another node variable PREV, which will always point to the node preceding NODE. When we reach the last node and its preceding node, that is, the second to last node, we will now change the next part of PREV to store the address of START. Hence, the memory occupied by the last node is freed. Finally, the second to last node now becomes the last node of the linked list. In this way, deletion of a node from the end is done in a circular linked list.

//Write a menu-driven program for circular linked lists performing insertion and deletion of all the cases.

```java
public class CircularLinkedList {
    private class Node {
        int data;
        Node next;
    }
    private Node start;
    private Node end;
    private int size;
    private Node create_new_node(int item) throws Exception {
        Node = new Node();
        if (node == null) {
            throw new Exception("Memory not allocated");
            } else {
            node.data = item;
            node.next = null;
```

```
            return node;
    }
}
public void insertion(int item, int idx) throws Exception {
    if (idx < 0 || idx > size) {
        throw new Exception("Invalid Index");
    }
    if (idx == 0) {
        insertion_at_beginning(item);
    } else if (idx == size) {
        insertion_at_end(item);
    } else {
        Node = create_new_node(item);
        Node prev = getNodeAt(idx - 1);
        node.next = prev.next;
        prev.next = node;
        end.next = start;
        size++;
    }
}
private void insertion_at_beginning(int item) throws Exception {
    Node = create_new_node(item);
    if (size == 0) {
        start = end = node;
    } else {
        node.next = start;
        start = node;
    }
    end.next = start;
    size++;
}
private void insertion_at_end(int item) throws Exception {
    Node = create_new_node(item);
    if (size > 0) {
        end.next = node;
        end = node;
    }
    if (size == 0) {
        start = end = node;
    }
    end.next = start;
    size++;
}
```

```java
public void deletion(int pos) throws Exception {
    if (pos < 0 || pos == size) {
        throw new Exception("Invalid Index");
    }
    if (size == 0) {
        throw new Exception("Linked List is Empty");
    }
    if (pos == 0) {
        deletion_at_beginning(pos);
    } else if (pos == size - 1) {
        deletion_at_end(pos);
    } else {
        Node prev = getNodeAt(pos - 1);
        Node next = getNodeAt(pos + 1);
        prev.next = next;
    }
    size--;
}
private void deletion_at_beginning(int pos) {
    end.next = start.next;
    start = start.next;
}
private void deletion_at_end(int pos) throws Exception {
    Node prev = getNodeAt(pos - 1);
    prev.next = start;
    end = prev;
}
public void display() throws Exception {
    if (size == 0) {
        throw new Exception("Linked List is Empty");
    }
    Node temp = start;
    while (temp != end) {
        System.out.print(temp.data + "->");
        temp = temp.next;
    }
    System.out.print(end.data + "->");
    System.out.print("NULL \n");
}
public Node getNodeAt(int idx) throws Exception {
    if (size == 0) {
        throw new Exception("Linked List is Empty");
    }
}
```

```
            if (idx < 0 || idx >= size) {
                throw new Exception("Invalid Index");
            }
            Node temp = start;
            for (int i = 0; i < idx; i++) {
                temp = temp.next;
            }
            return temp;
        }
    }
}
//CLIENT CLASS
import java.util.Scanner;
public class CLLClient {
    public static void main(String[] args) {
        Scanner scn = new Scanner(System.in);
        CircularLinkedList list = new CircularLinkedList();
        boolean flag = true;
        try {
            while (flag) {
                System.out.println("\n ***MENU***");
                System.out.println("1. Insertion");
                System.out.println("2. Deletion");
                System.out.println("3. Display");
                System.out.println("4. Exit");
                System.out.println("Enter your choice: ");
                int choice = scn.nextInt();
                int item = 0, pos = 0;
                switch (choice) {
                case 1:
                    System.out.println("Enter value of node: ");
                    item = scn.nextInt();
                    System.out.println("Enter position of
                                    node:");
                    pos = scn.nextInt();
                    list.insertion(item, pos - 1);
                    System.out.println(item + " inserted
                                    successfully at " + pos);
                    break;
                case 2:
                    System.out.println("Enter position of
                                    node:");
                    pos = scn.nextInt();
                    list.deletion(pos - 1);
                    System.out.println("Item deleted
                                    successfully");
```

```
                                break;
                        case 3:
                                list.display();
                                break;
                        case 4:
                                flag = false;
        System.out.println("Terminated.....");
                                break;
                        default:
                                System.out.println("Wrong choice");
                                break;
            }

                }
        } catch (Exception e) {
            System.out.println(e.getMessage());
        }
    }
}
```

The output of the program is shown as:

```
40 inserted successfully at 4

***MENU***
1. Insertion
2. Deletion
3. Display
4. Exit
Enter your choice:
1
Enter value of node:
50
Enter position of node:
5
50 inserted successfully at 5

***MENU***
1. Insertion
2. Deletion
3. Display
4. Exit
Enter your choice:
3
10->20->30->40->50->NULL

***MENU***
1. Insertion
2. Deletion
3. Display
4. Exit
Enter your choice:
2
Enter position of node:
3
Item deleted successfully

***MENU***
1. Insertion
2. Deletion
3. Display
4. Exit
Enter your choice:
2
Enter position of node:
1
Item deleted successfully

***MENU***
1. Insertion
2. Deletion
3. Display
4. Exit
Enter your choice:
3
20->40->50->NULL

***MENU***
1. Insertion
2. Deletion
3. Display
4. Exit
Enter your choice:
4
Terminated.....

C:\Program Files\Java\jdk-12.0.2\bin>
```

4.4.5 Doubly Linked List

A doubly linked list is also called a two-way linked list; it is a special type of linked list which can point to the next node as well as the previous node in the sequence. In a doubly linked list, each node is divided into three parts:

1. The first part is called the previous node, which contains the address of the previous node in the list.

2. The second part is called the information part, which contains the information of the node.

3. The third part is called the next node, which contains the address of the succeeding node in the list.

START   END

| NULL | 8 | | | | 6 | | | | 9 | | | | 2 | NULL |

prev   data   next

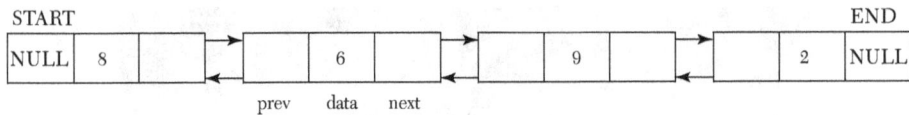

Figure 4.15. Doubly linked list.

The structure of a doubly linked list is given as follows:

```
private class Node {
      Node prev;
      int data;
      Node next;
}
```

The first node of the linked list will contain a NULL value in the previous node to indicate that there is no element preceding it in the list; similarly, the last node will also contain a NULL value in the next node field to indicate that there is no element succeeding it in the list. Doubly linked lists can be traversed in both directions.

4.4.6   Operations on a Doubly Linked List

Various operations can be performed on a circular linked list, which include the following:

- Inserting a New Node in a Doubly Linked List

- Deleting a Node from a Doubly Linked List

Let us now discuss both these operations in detail.

(a) Inserting a New Node in a Doubly Linked List

In this section, we will learn how a new node is inserted into an already existing doubly linked list. We will consider four cases for the insertion process in a doubly linked list.

1. A new node is inserted at the beginning.

2. A new node is inserted at the end.

3. A new node is inserted after a given node.

4. A new node is inserted before a given node.

1. Inserting a new node in the beginning of a doubly linked list

In this case of inserting a new node in the beginning of a doubly linked list, we will first check with the overflow condition, that is, whether the memory is available for a new node. If the memory is not available, then an overflow message is displayed; otherwise, the memory is allocated for the new node. Now, we will initialize the node with its info part, and its address part will contain the address of the first node of the list, which is the START node. Hence, the new node is added as the first node in the list, and the START node will point to the first node of the list. Now, to understand better, let us take an example. Consider a linked list as shown in the following figure with four nodes; a new node will be inserted in the beginning of the linked list.

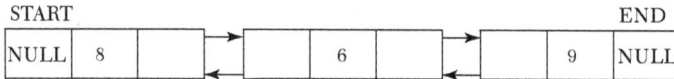

To meet 2 at the beginning of the Doubly Linked List, first create a newNode with newNode.prev=NULL newNode.data=2 and newNode.next=NULL.

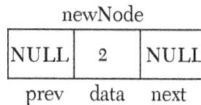

Now assign newNode.next=START and START=newNode just the same way we did in Linked List. Now START contains 2 instead of 8. This way we can insert a node at the beginning of Doubly Linked List.

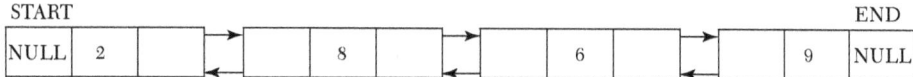

Figure 4.16. Inserting a new node in the beginning of a doubly linked list.

From the previous example, it is clear how a new node will be inserted in an already existing doubly linked list. Let us now understand its algorithm:

Algorithm for inserting a new node in the beginning of a doubly linked list

```
Step 1: START
Step 2: IF NEW NODE = NULL
            Print OVERFLOW
        [End of If]
Step 3: Set NEW NODE. INFO = VALUE
Step 4: Set NEW NODE. PREV = NULL
Step 5: Set NEW NODE. NEXT = START
Step 7: Set START. PREV = NEW NODE
Step 8: Set START = NEW NODE
Step 9: EXIT
```

2. Inserting a new node at the end of a doubly linked list

In the case of inserting the new node at the end of the linked list, we will first check the overflow condition, which is whether the memory is available for a new node. If the memory is not available, then an overflow message is printed; otherwise, the memory is allocated for the new node.

Then a NODE variable is made which will initially point to START, and a NODE variable will be used to traverse the list until it reaches the last node. When it reaches the last node, the NEXT part of the last node will store the address of the new node, and the NEXT part of the NEW NODE will contain NULL, which will denote the end of the linked list. The PREV part of the NEW NODE will store the address of the node pointed to by NODE. Let us understand it with the help of an algorithm:

Algorithm for inserting a new node at the end of a linked list

```
Step 1: START
Step 2: IF NODE = NULL
            Print OVERFLOW
        [End of If]
Step 3: Set NEW NODE. INFO = VALUE
Step 4: Set NEW NODE. NEXT = NULL
Step 5: Set NEW NODE. NEXT = NULL
Step 6: Set TEMP = START
Step 8: Repeat while TEMP. NEXT != NULL
            Set TEMP = TEMP. NEXT
        [End of Loop]
Step 9: Set TEMP. NEXT = NEW NODE
Step 10: Set NEW NODE. PREV = TEMP
Step 11: EXIT
```

From the previous algorithm we understand how to insert a new node at the end of a doubly linked list. Now, we will study this further with the help of an example. Consider a linked list as shown in the following figure with four nodes; a new node will be inserted at the end of the doubly linked list:

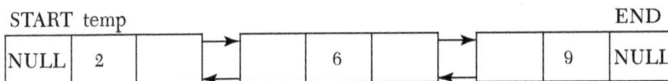

To meet 8 at the end of the Doubly Linked List, first create a newNode with newNode.prev=NULL, newNode.data=8 and newNode.next=NULL. Also declare a node temp=START. This temp will traverse the Doubly Linked List till temp.next!=NULL.

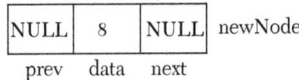

When temp.next is equal to NULL assign temp.next=newNode and newNode.prev=temp.

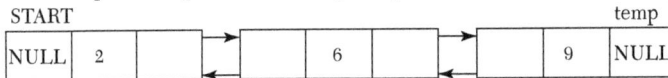

Or you can just maintain a node END that represents the last node and assign END.next-newNode and newNode.pre=END, then END=newNode. This way you can insert a node at the end of a Doubly Linked List.

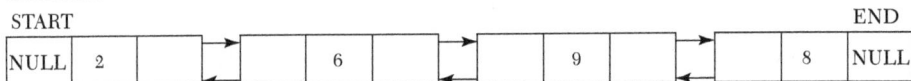

Figure 4.17. Inserting a new node at the end of a doubly linked list.

3. Inserting a new node after a given node in a doubly linked list

In this case, a new node is inserted after a given node in a doubly linked list. As in the other cases, we will again check the overflow condition in it. If the memory for the new node is available, then it will be allocated; otherwise, an overflow message is displayed. Then a NODE variable is made which will initially point to START, and the NODE variable is used to traverse the linked list until its value becomes equal to the value after which the new node is to be inserted. When it reaches that node/value, then the NEXT part of that node will store the address of the new node, and the PREV part of the NEW NODE will store the address of the preceding node. Let us understand it with the help of the following algorithm:

Algorithm for inserting a new node after a given node in a linked list

```
Step 1:  START
Step 2:  IF NODE = NULL
              Print OVERFLOW
         [End of If]
Step 3:  Set NEW NODE. INFO = VALUE
Step 4:  Set TEMP = START
Step 5:  Repeat while TEMP. INFO != GIVEN_VAL
              Set TEMP = TEMP. NEXT
         [End of Loop]
Step 6:  Set TEMP2 = TEMP.NEXT
Step 7:  Set NEW NODE. NEXT = TEMP. NEXT
Step 8:  Set NEW NODE. PREV = NODE
Step 9:  Set TEMP. NEXT = NEW NODE
Step 10: Set TEMP2. PREV = NEW NODE
Step 11: EXIT
```

Now, we will understand more about the same with the help of an example. Consider a doubly linked list as shown in the following figure with four nodes; a new node will be inserted after a given node in the linked list:

START temp   END

| NULL | 2 | | | 6 | | | 9 | NULL |

To meet 8 in between the Doubly Linked List after 6, first create a newNode with
newNode.prev=NULL, newNode.data=8 and newNode.next=NULL. Also declare a node
temp=START that will traverse the Doubly Linked List till temp.data is equal to 6.

| NULL | 8 | NULL | newNode

prev data next

START   temp   temp2 END

| NULL | 2 | | | 6 | | | 9 | NULL |

Once temp.data is equal to 6, assign newNode.prev=temp, newNode.next=temp.next. The declare a node
temp2=temp.next. Assign temp2.prev=newNode and temp.next=newNode. This way we can insert a node
after a node in Doubly Linked List.

START   END

| NULL | 2 | | | 6 | | | 8 | | | 9 | NULL |

Figure 4.18. Inserting a new node after a given node in a doubly linked list.

4. Inserting a new node before a given node in a doubly linked list

In this case, a new node is inserted before a given node in a doubly linked
list. As in the other cases, we will again check the overflow condition in
it. If the memory for the new node is available, then it will be allocated;
otherwise, an overflow message is displayed. Then a NODE variable is
made which will initially point to START, and the NODE variable is used
to traverse the linked list until its value becomes equal to the value before

START temp   END

| NULL | 2 | | | 8 | | | 9 | NULL |

To meet a node 6 before 8 in Doubly Linked List, first create a newNode with
newNode.prev=NULL, newNode.data=6 and newNode.next=NULL. Then declare a node
temp=START that will traverse the Doubly Linked List till temp.data is equal to 8.

| NULL | 6 | NULL | newNode

prev data next

START temp2   temp   END

| NULL | 2 | | | 8 | | | 9 | NULL |

Once temp.data is equal to 8, assign newNode.next=temp, newNode.prev=temp.prev. The declare a node
temp2=temp.prev. Assign temp2.next=newNode and temp.prev=newNode. This way we can insert a node
before a node in Doubly Linked List.

START   END

| NULL | 2 | | | 6 | | | 8 | | | 9 | NULL |

Figure 4.19. Inserting a new node before a given node in a doubly linked list.

which the new node is to be inserted. When it reaches that node/value, then the PREV part of that node will store the address of the NEW NODE, and the NEXT part of the NEW NODE will store the address of the succeeding node. Now, to understand better, let us take an example. Consider a linked list as shown in the following figure with four nodes; a new node will be inserted before a given node in the linked list.

From the previous example, it is clear how a new node will be inserted in an already existing doubly linked list. Let us now understand its algorithm:

Algorithm for inserting a new node before a given node in a doubly linked list

```
Step 1: START
Step 2: IF NEW NODE = NULL
            Go to Step 10
        [End of If]
Step 3: Set NEW NODE. INFO = VALUE
Step 4: Set TEMP = START
Step 5: Repeat while TEMP. INFO != GIVEN_VAL
            Set TEMP = TEMP. NEXT
        [End of Loop]
Step 6: Set TEMP2 = TEMP. PREV
Step 7: Set NEW NODE. NEXT = TEMP
Step 8: Set NEW NODE. PREV = TEMP2
Step 9: Set TEMP. PREV = NEW NODE
Step 10: Set TEMP2. NEXT = NEW NODE
Step 11: EXIT
```

(b) Deleting a Node from a Doubly Linked List

In this section, we will learn how a node is deleted from an already existing doubly linked list. We will consider four cases for the deletion process in a doubly linked list.

1. A node is deleted from the beginning of the linked list.

2. A node is deleted from the end of the linked list.

3. A node is deleted after a given node from the linked list.

4. A node is deleted before a given node from the linked list.

Now let us discuss the previous cases in detail.

1. Deleting a node from the beginning of the doubly linked list

In the case of deleting a node from the beginning of the doubly linked list, we will first check the underflow condition, which occurs when we try to delete a node from the linked list which is empty. This situation exists when the START node is equal to NULL. Hence, if the condition is true, then the underflow message is displayed; otherwise, the node is deleted from the linked list. Consider a linked list as shown in the following figure with five nodes; the node will be deleted from the beginning of the linked list.

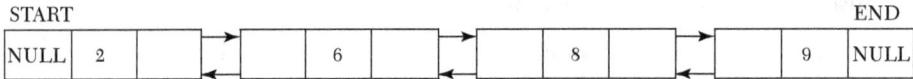

START   END

| NULL | 2 | | | 6 | | | 8 | | | 9 | NULL |

To delete a node from the beginning of a Doubly Linked List, assin START=START.next and then START.prev=NULL. This way 6 becomes the START and the first mode is deleted from the Doubly Linked ListS.

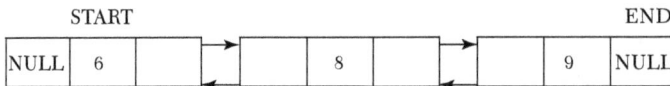

START   END

| NULL | 6 | | | 8 | | | 9 | NULL |

Figure 4.20. Deleting a node from the beginning of the doubly linked list.

Let us understand this with the help of an algorithm:

Algorithm for deleting a node from the beginning of a doubly linked list

```
Step 1: START
Step 2: IF START = NULL
            Print UNDERFLOW
        [End Of If]
Step 3: Set START = START. NEXT
Step 4: Set START. PREV = NULL
Step 5: EXIT
```

In the previous algorithm, first we check for the underflow condition, which is whether there are any nodes present in the linked list or not. If there are no nodes, then an underflow message will be printed. Otherwise, we move to Step 3, where we are initialing NODE to START; that is, NODE will now store the address of the first node. In the next step START is moved to the second node, as now START will store the address of the second node. Also, the PREV part of the second node will now contain a value NULL. Hence, the first node is deleted and the memory that occupied NODE is freed (initially the first node of the list).

2. Deleting a node from the end of a doubly linked list

In the case of deleting a node from the end of a linked list, we will first check the underflow condition. This situation exists when the START node is equal to NULL. Hence, if the condition is true, then the underflow message is printed on the screen; otherwise, the node is deleted from the linked list. Consider a linked list as shown in the following figure with five nodes; the node will be deleted from the end of the linked list.

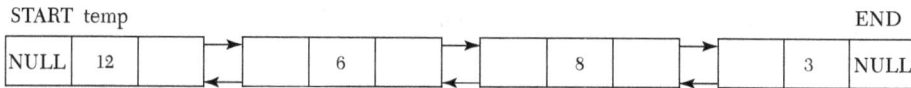

START temp   END

| NULL | 12 | | | 6 | | | 8 | | | 3 | NULL |

To delete a node from the end of a Doubly Linked List, declare a node temp=START that will traverse the Doubly Linked List till temp.next!=NULL. Once temp.next is equal to NULL declare another node temp2=temp.prev. Then assign temp2.next=NULL. This way the last node will become unreachable and will be deleted from the end of the Doubly Linked List.

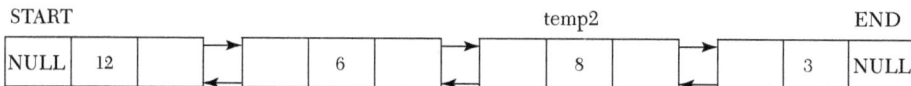

START   temp2   END

| NULL | 12 | | | 6 | | | 8 | | | 3 | NULL |

You can also maintain a node END that will keep record of the last node and declare a node temp2=End.prev. Then assign temp2.next=NULL and END=temp2.

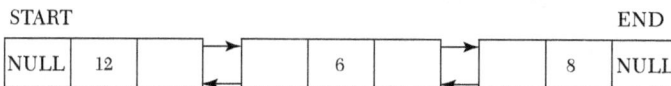

START   END

| NULL | 12 | | | 6 | | | 8 | NULL |

Figure 4.21. Deleting a node from the end of the doubly linked list.

From the previous example, it is clear how a node will be deleted from an already existing doubly linked list. Let us now understand its algorithm:

Algorithm for deleting a node from the end in a doubly linked list

```
Step 1: START
Step 2: IF START = NULL
            Print UNDERFLOW
        [End Of If]
Step 3: Set TEMP = START
Step 4: Repeat while TEMP. NEXT != NULL
            Set TEMP = TEMP. NEXT
        [End of Loop]
Step 5: Set TEMP2 = TEMP. PREV
Step 6: Set TEMP2. NEXT = NULL
Step 7: EXIT
```

In the previous algorithm, again we are checking for the underflow condition. If the condition is true, then the underflow message is printed. Otherwise, NODE is initialized to the START node; that is, NODE is pointing to the first node of the list. In the loop NODE is traversed until it reaches the last node of the list. After reaching the last node of the list, we can also access the second to last node by taking the address from the PREV part of the last node. Therefore, the last node is deleted, and the memory that occupied NODE is now freed.

3. Deleting a node after a given node from the doubly linked list

In the case of deleting a node after a given node from the linked list, we will again check the underflow condition as we checked in both the other cases. This situation exists when the START node is equal to NULL. Hence, if the condition is true, then the underflow message is displayed; otherwise, the node is deleted from the linked list. Consider a linked list as shown in the following figure with five nodes; the node will be deleted after a given node from the linked list.

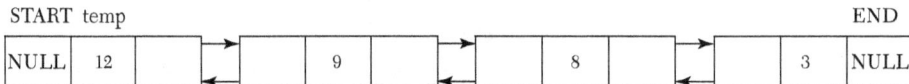

To delete a node after a specific node in a Doubly Linked List, first declare a node temp=START. So to delete 8, this temp will traverse the Doubly Linked List till temp.data!=9. When temp.data is equal to 9, declare another node temp2=temp.next.next
The mode between temp and temp2 is to be deleted.

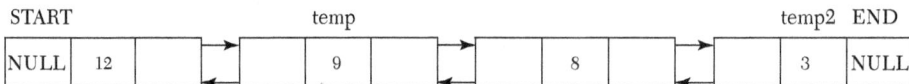

Now assign temp.next=temp2 and temp2.prev=temp. Hence, the node 8 will become unreachable. This way you can delete a node present after a node in a Doubly Linked List.

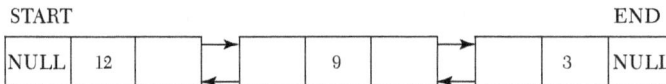

Figure 4.22. Deleting a node after a given node from the doubly linked list.

Now let us understand the previous case with the help of an algorithm.

Algorithm for deleting a node after a given node from the linked list

```
Step 1: START
Step 2: IF START = NULL
            Print UNDERFLOW
        [End Of If]
Step 3: Set TEMP = START
Step 4: Repeat while TEMP. INFO != GIVEN_VAL
            Set TEMP = TEMP. NEXT
        [End of Loop]
Step 5: Set TEMP2 = TEMP. NEXT. NEXT
Step 6: Set TEMP. NEXT = TEMP2
Step 7: Set TEMP2. PREV = TEMP
Step 8: EXIT
```

In the previous algorithm, first we check for the underflow condition. If the condition is true, then the underflow message is printed. Otherwise, NODE is initialized to the START node; that is, NODE is pointing to the first node of the list. In the loop NODE is moved until its info part becomes equal to the node after which the node is to be deleted. After reaching that node of the list, we can also access the succeeding node by taking the address from the NEXT part of that node. Therefore, the node is deleted, and the memory which was being occupied by the TEMP is now free.

4. Deleting a node before a given node from the doubly linked list

In the case of deleting a node before a given node from the linked list, we will again check the underflow condition as we checked in both the other cases. This situation occurs when the START node is equal to NULL. Hence, if the condition is true, then the underflow message is printed; otherwise, the node is deleted from the linked list. Consider a linked list as shown in the following figure with five nodes; the node will be deleted before a given node from the linked list.

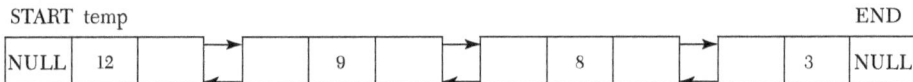

To delete a node before a node in a Doubly Linked List, first declare a node temp=START. So to delete 9, this temp will traverse the Doubly Linked List till temp!=8. Once temp.data is equal to 8, then declare another node temp2=temp.prev.prev
The mode between temp2 and temp is to be deleted.

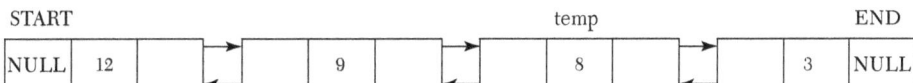

Assign temp2.next=temp and temp.prev=temp2. Hence, node 9 will become unreachable. This way you can delete a node present before a node in a Doubly Linked List.

Figure 4.23. Deleting a node before a given node from the doubly linked list.

From the previous example, it is clear how a node will be deleted from an already existing doubly linked list. Let us now understand its algorithm:

Algorithm for deleting a node before a given node in a doubly linked list

```
Step 1: START
Step 2: IF START = NULL
            Print UNDERFLOW
        [End Of If]
Step 3: Set TEMP = START
Step 4: Repeat while TEMP. INFO != GIVEN_VAL
            Set TEMP = TEMP. NEXT
        [End of Loop]
```

Step 5: Set TEMP2 = TEMP. PREV. PREV
Step 6: Set TEMP. PREV = TEMP2
Step 7: Set TEMP2. NEXT = TEMP
Step 10: EXIT

In the previous algorithm, first we are checking for the underflow condition. If the condition is true, then the underflow message is printed. Otherwise, NODE is initialized to the START node; that is, NODE is pointing to the first node of the list. In the loop NODE is moved until its info part becomes equal to the node before which the node is to be deleted. After reaching that node of the list, we can also access the preceding node by taking the address from the PREV part of that node. Therefore, the node is deleted, and the memory which was being occupied by the TEMP is now free.

//Write a menu-driven program for doubly linked lists, performing insertion and deletion of all cases.

```java
public class DoublyLinkedList {
    private class Node {
        Node prev;
        int data;
        Node next;
    }
    private Node start;
    private int size;
    private Node create_new_node(int item) throws Exception {
        Node = new Node();
        if (node == null) {
            throw new Exception("Memory not allocated");
        } else {
            node.data = item;
            node.prev = null;
            node.next = null;
            return node;
        }
    }
    public void insertion(int item, int pos) throws Exception {
        if (pos < 0 || pos > size) {
            throw new Exception("Invalid Index");
        }
        if (pos == 0) {
            insertion_at_beginning(item);
        } else if (pos == size) {
            insertion_at_end(item);
```

```
        } else {
            Node = create_new_node(item);
            Node preNode = getNodeAt(pos - 1);
            Node nextNode = preNode.next;
            node.next = nextNode;
            node.prev = preNode;
            nextNode.prev = node;
            preNode.next = node;
            size++;
        }
    }
    private void insertion_at_beginning(int item) throws Exception {
        Node = create_new_node(item);
        Node temp = new Node();
        if (start == null) {
            start = node;
        } else {
            temp = start;
            start = node;
            start.next = temp;
        }
        size++;
    }
    private void insertion_at_end(int item) throws Exception {
        Node = create_new_node(item);
        Node temp = new Node();
        temp = start;
        while (temp.next != null) {
            temp = temp.next;
        }
        temp.next = node;
        node.prev = temp;
        size++;
    }
    public void deletion(int pos) throws Exception {
        if (pos < 0 || pos == size) {
            throw new Exception("Invalid Index");
        }
        if (size == 0) {
            throw new Exception("Linked List is Empty");
        }
        if (pos == 0) {
            deletion_at_beginning();
```

```java
        } else if (pos == size - 1) {
            deletion_at_end(pos);
        } else {
            Node preNode = getNodeAt(pos - 1);
            Node nextNode = getNodeAt(pos + 1);
            preNode.next = nextNode;
            nextNode.prev = preNode;
        }
        size--;
    }
    private void deletion_at_beginning() {
        start = start.next;
        start.prev = null;
    }
    private void deletion_at_end(int pos) throws Exception {
        Node prev = getNodeAt(pos - 1);
        prev.next = null;
    }
    public void display() throws Exception {
        if (size == 0) {

throw new Exception("Linked List is Empty");
        }
        Node temp = start;
        while (temp != null) {
            System.out.print(temp.data + "->");
            temp = temp.next;
        }
        System.out.print("NULL \n");
    }
    public Node getNodeAt(int idx) throws Exception {
        if (size == 0) {
            throw new Exception("Linked List is Empty");
        }
        if (idx < 0 || idx >= size) {
            throw new Exception("Invalid Index");
        }
        Node temp = start;
        for (int i = 0; i < idx; i++) {
            temp = temp.next;
        }
        return temp;
    }
}
```

```java
//CLIENT CLASS
import java.util.Scanner;
public class DLLClient {
    public static void main(String[] args) {
        Scanner scn = new Scanner(System.in);
        DoublyLinkedList list = new DoublyLinkedList();
        boolean flag = true;
        try {
            while (flag) {
                System.out.println("\n***MENU***");
                System.out.println("1. Insertion");
                System.out.println("2. Deletion");
                System.out.println("3. Display");
                System.out.println("4. Exit");
                System.out.println("Enter your choice: ");
                int choice = scn.nextInt();
                int item = 0, pos = 0;
                switch (choice) {
                case 1:
                    System.out.println("Enter value of node: ");
                    item = scn.nextInt();
                    System.out.println("Enter position of
                                        node:");
                    pos = scn.nextInt();
                    list.insertion(item, pos - 1);
                    System.out.println(item + " inserted
                                        successfully at " + pos);
                    break;
                case 2:
                    System.out.println("Enter position of
                                        node:");
                    pos = scn.nextInt();
                    list.deletion(pos - 1);
                    System.out.println("Item deleted
                                        successfully");
                    break;
                case 3:
                    list.display();
                    break;
                case 4:
                    flag = false;
                    System.out.println("Terminated.....");
                    break;
```

```
                default:
                        System.out.println("Wrong choice");
                        break;
                }
        }
    } catch (Exception e) {
        System.out.println(e.getMessage());
    }

}
```

The output of the program is shown as:

Now let us discuss header linked lists.

4.5   Header Linked Lists

Header linked lists are a special type of linked list which always contain a special node, called the header node, at the beginning. This header node usually contains vital information about the linked list, like the total number of nodes in the list, whether the list is sorted or not, and so on. There are two types of header linked lists, which include the following:

1. Grounded Header Linked List: This linked list stores a unique value NULL in the address field (next part) of the last node of the list.

Header Node

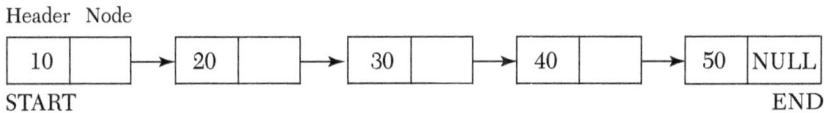

START   END

Figure 4.24. Grounded header linked list.

2. Circular Header Linked List: This linked list stores the address of the header node in the address field (next part) of the last node of the list.

Header Node   END

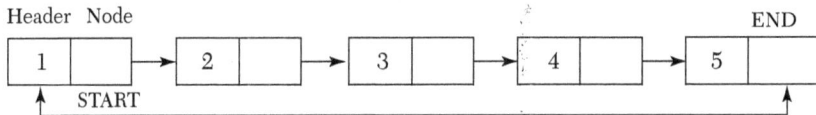

Figure 4.25. Circular header linked list.

Frequently Asked Questions

3. What are the uses of a header node in a linked list?

Ans: *The header node is a node of a linked list which may or may not have the same data structure of that of a typical node. The only common thing between a typical node and a header node is that they both are referring to a typical node.*

//Write a program to implement a header linked list.

```
public class HeaderLinkedList {
    private class Node {
        int data;
        Node next;
    }
    private Node start;
    private Node header;
    private int size;
    private Node create_new_node(int item) throws Exception {
        Node node = new Node();
```

```java
        if (node == null) {
            throw new Exception("Memory not allocated");
        } else {
            node.data = item;
            node.next = null;
            return node;
        }
    }
    public void insertion(int item, int pos) throws Exception {
        if (pos < 0 || pos > size) {
            throw new Exception("Invalid Index");
        }
        if (pos == 0) {
            insertion_at_beginning(item);
        } else if (pos == size) {
            insertion_at_end(item);
        } else {
            Node node = create_new_node(item);
            Node prev = getNodeAt(pos - 1);
            node.next = prev.next;
            prev.next = node;
            size++;
        }
    }
    private void insertion_at_beginning(int item) throws Exception {
        Node node = create_new_node(item);
        Node temp = new Node();
        if (start.next == null) {
            header.next = node;
            node.next = null;
        } else {
            temp = start.next;
            header.next = node;
            node.next = temp;
        }
        size++;
    }
    private void insertion_at_end(int item) throws Exception {
        Node node = create_new_node(item);
        Node temp = new Node();
        temp = header;
        while (temp.next != null) {
            temp = temp.next;
        }
```

```
            temp.next = node;
            size++;
    }
    public void deletion(int pos) throws Exception {
            if (pos < 0 || pos == size) {
                throw new Exception("Invalid Index");
            }
            if (size == 0) {
                throw new Exception("Linked List is Empty");
            }
            if (pos == 0) {
                deletion_at_beginning();
            } else if (pos == size - 1) {
                deletion_at_end(pos);
            } else {
                Node prev = getNodeAt(pos - 1);
                Node next = getNodeAt(pos + 1);
                prev.next = next;
            }
            size--;
    }
    private void deletion_at_beginning() {
            Node temp = start.next;
            header.next = temp.next;
    }
    private void deletion_at_end(int pos) throws Exception {
            Node prev = getNodeAt(pos - 1);
            prev.next = null;
    }
    public void display() throws Exception {
            if (size == 0) {
                throw new Exception("Linked List is Empty");
            }
            Node temp = header.next;
            while (temp != null) {
                System.out.print(temp.data + "->");
                temp = temp.next;
            }
            System.out.print("NULL \n");
    }
    public Node getNodeAt(int idx) throws Exception {
            if (size == 0) {
                throw new Exception("Linked List is Empty");
            }
```

```java
        if (idx < 0 || idx >= size) {
            throw new Exception("Invalid Index");
        }
        Node temp = header;
        for (int i = 0; i < idx; i++) {
            temp = temp.next;
        }
        return temp;
    }
}
//CLIENT CLASS
import java.util.Scanner;
public class HLLClient {
    public static void main(String[] args) {
        Scanner scn = new Scanner(System.in);
        HeaderLinkedList list = new HeaderLinkedList();
        boolean flag = true;
        try {
            while (flag) {
                System.out.println("\n***MENU***");
                System.out.println("1. Insertion");
                System.out.println("2. Deletion");
                System.out.println("3. Display");
                System.out.println("4. Exit");
                System.out.println("Enter your choice: ");
                int choice = scn.nextInt();
                int item = 0, pos = 0;
                switch (choice) {
                case 1:
                    System.out.println("Enter value of node: ");
                    item = scn.nextInt();
                    System.out.println("Enter position of
                                        node:");
                    pos = scn.nextInt();
                    list.insertion(item, pos - 1);
                    System.out.println(item + " inserted
                                        successfully at " + pos);
                    break;
                case 2:
                    System.out.println("Enter position of
                                        node:");
                    pos = scn.nextInt();
                    list.deletion(pos - 1);
```

```java
                System.out.println("Item deleted
                                    successfully");
                break;
            case 3:
                list.display();
                break;
            case 4:
                flag = false;
                System.out.println("Terminated.....");
                break;
                default:
                System.out.println("Wrong choice");
                break;
            }
        }
    } catch (Exception e) {
        System.out.println(e.getMessage());
    }
  }
}
```

The output of the program is shown as:

//Write a program to implement a circular header linked list.

```java
public class CircularHeaderLinkedList {
    private class Node {
        int data;
        Node next;
    }
    private Node start;
    private Node header;
    private Node end;
    private int size = -1;
    private Node create_new_node(int item) throws Exception {
        Node node = new Node();
        if (node == null) {
            throw new Exception("Memory not allocated");
        } else {
            node.data = item;
            node.next = null;
            return node;
        }
    }
    public void insertion(int item, int idx) throws Exception {
        if (idx < 0 || idx > size) {
            throw new Exception("Invalid Index");
        }
        if (idx == 0) {
            insertion_at_beginning(item);
        } else if (idx == size) {
            insertion_at_end(item);
        } else {
            Node node = create_new_node(item);
            Node prev = getNodeAt(idx - 1);
            node.next = prev.next;
            prev.next = node;
```

```
            end.next = start;
            size++;
        }
    }
    private void insertion_at_beginning(int item) throws Exception {
        Node node = create_new_node(item);
        if (size == 0) {
            start.next = node;
            header = start;
            end = node;
        } else {
            Node temp = start.next;
            node.next = temp;
            header.next = node;
        }
        end.next = start;
        size++;
    }
    private void insertion_at_end(int item) throws Exception {
        Node node = create_new_node(item);
        if (size > 0) {
            end.next = node;
            end = node;
        }
        if (size == 0) {
            start = header = end = node;
        }
        end.next = start;
        size++;
    }
    public void deletion(int pos) throws Exception {
        if (pos < 0 || pos == size) {
            throw new Exception("Invalid Index");
        }
        if (size == 0) {
            throw new Exception("Linked List is Empty");
        }
        if (pos == 0) {
            deletion_at_beginning(pos);
        } else if (pos == size - 1) {
            deletion_at_end(pos);
        } else {
            Node prev = getNodeAt(pos - 1);
            Node next = getNodeAt(pos + 1);
            prev.next = next;
```

```
        }
        size--;
    }
    private void deletion_at_beginning(int pos) {
        Node temp =start.next;
        header.next = temp.next;
    }
    private void deletion_at_end(int pos) throws Exception {
        Node prev = getNodeAt(pos - 1);
        prev.next = start;
        end = prev;
    }
    public void display() throws Exception {
        if (size == 0) {
            throw new Exception("Linked List is Empty");
        }
        Node temp = header.next;
        while (temp != end) {
            System.out.print(temp.data + "->");
            temp = temp.next;
        }
        System.out.print(end.data + "->");
        System.out.print("NULL \n");
    }
    public Node getNodeAt(int idx) throws Exception {
        if (size == 0) {
            throw new Exception("Linked List is Empty");
        }
        if (idx < 0 || idx >= size) {
            throw new Exception("Invalid Index");
        }
        Node temp = header;
        for (int i = 0; i < idx; i++) {
            temp = temp.next;
        }
        return temp;
    }
}
//CLIENT CLASS
import java.util.Scanner;
public class CHLLClient {
    public static void main(String[] args) {
        Scanner scn = new Scanner(System.in);
```

```java
CircularHeaderLinkedList list = new
                            CircularHeaderLinkedList();
boolean flag = true;
try {
    while (flag) {
        System.out.println("\n***MENU***");
        System.out.println("1. Insertion");
        System.out.println("2. Deletion");
        System.out.println("3. Display");
        System.out.println("4. Exit");
        System.out.println("Enter your choice: ");
        int choice = scn.nextInt();
        int item = 0, pos = 0;
        switch (choice) {
        case 1:
            System.out.println("Enter value of node: ");
            item = scn.nextInt();
            System.out.println("Enter position of node:");
            pos = scn.nextInt();
            list.insertion(item, pos - 1);
            System.out.println(item + " inserted
                            successfully at " + pos);
            break;
        case 2:
            System.out.println("Enter position of
                                node:");
            pos = scn.nextInt();
            list.deletion(pos - 1);
            System.out.println("Item deleted
                                successfully");
            break;
        case 3:
            list.display();
            break;
        case 4:
            flag = false;
            System.out.println("Terminated.....");
            break;
        default:
            System.out.println("Wrong choice");
            break;
        }
```

```
        }
    } catch (Exception e) {
        System.out.println(e.getMessage());
        }
    }
}
```

The output of the program is shown as:

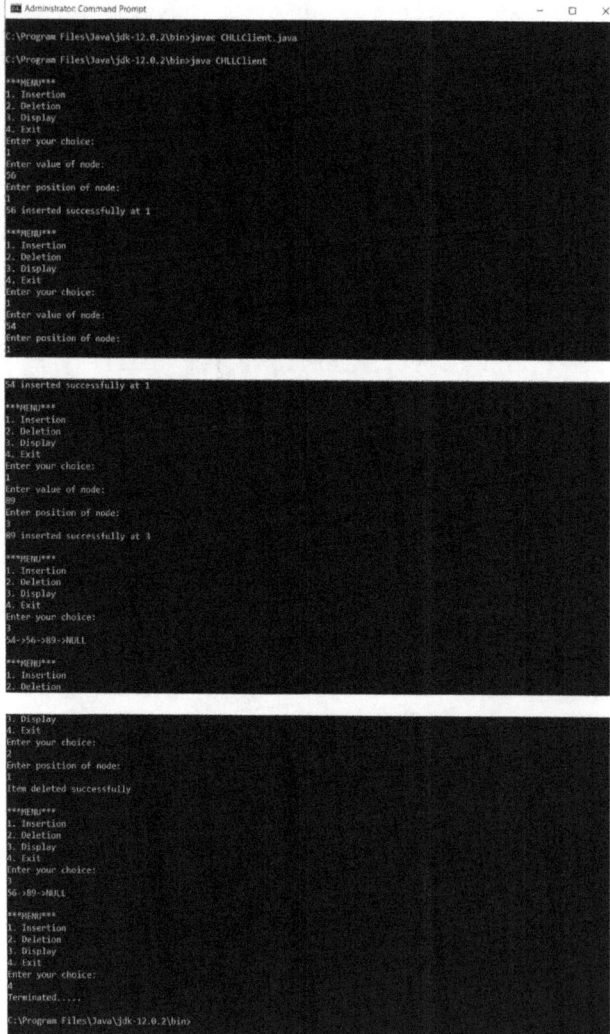

4.6   Applications of Linked Lists

Linked lists have various applications, but one of the most important applications of linked lists is polynomial representation; linked lists can be used to represent polynomials, and there are different operations that can be performed on them. Now let us see how polynomials can be represented in the memory using linked lists.

4.7   Polynomial Representation

Consider a polynomial $10x^2 + 6x + 9$. In this polynomial, every individual term consists of two parts: first, a coefficient, and second, a power. Here, the coefficients of the expression are 10, 6, and 9, and 2, 1, and 0 are the respective powers of the coefficients. Now, every individual term can be represented using a node of the linked list. The following figure shows how a polynomial expression can be represented using a linked list:

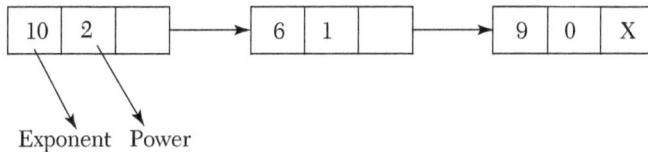

Figure 4.26. Linked representation of a polynomial.

4.8   Summary

- A linked list is a sequence of nodes in which each node contains one or more than one data field and a node which points to the next node.

- The process of allocating memory during the execution of the program or the process of allocating memory to the variables at runtime is called dynamic memory allocation.

- A singly linked list is the simplest type of linked list, in which each node contains some information/data and only one node which points to the next node in the linked list.

- Traversing a linked list means accessing all the nodes of the linked list exactly once.

- Searching for a value in a linked list means to find a particular element/value in the linked list.

- A circular linked list is also a type of singly linked list in which the address part of the last node will store the address of the first node.

▪ A doubly linked list is also called a two-way linked list; it is a special type of linked list which can point to the next node as well as the previous node in the sequence.

▪ A header linked list is a special type of linked list which always contains a special node, called the header node, at the beginning. This header node usually contains vital information about the linked list like the total number of nodes in the list, whether the list is sorted or not, and so forth.

▪ One of the most important applications of linked lists is polynomial representation, because linked lists can be used to represent polynomials, and there are different operations that can be performed on them.

4.9 Exercises

4.9.1 Theory Questions

1. What is a linked list? How it is different from an array?

2. How many types of linked lists are there? Explain in detail.

3. What is the difference between singly and doubly linked lists?

4. List the various advantages of linked lists over arrays.

5. What is a circular linked list? What are the advantages of a circular linked list over a linked list?

6. Define a header linked list and explain its utility.

7. Give the linked representation of the following polynomial: $10x^2y - 6x + 7$.

8. Specify the use of a header node in a header linked list.

9. List the various operations that can be performed in linked lists.

4.9.2 Programming Questions

1. Write an algorithm/program to insert a node at a desired position in a circular linked list.

2. Write a Java program to insert and delete the node at the beginning in a doubly linked list using classes.

3. Write an algorithm to reverse a singly linked list.

4. Write a Java program to delete a node from a header linked list.

5. Write an algorithm to concatenate two linked lists.

6. Write a Java program to implement a circular header linked list.

7. Write a Java program to count the non-zero values in a header linked list using classes.

8. Write a Java program that inserts a node in the linked list before a given node.

9. Write an algorithm to search for an element from a given linear linked list.

10. Write a program that inserts a node in a doubly linked list after a given node.

4.9.3 Multiple Choice Questions

1. Linked lists are best suited for:
- **(a)** Data structure
- **(b)** Size of structure and data are constantly changing
- **(c)** Size of structure and data are fixed
- **(d)** None of these

2. Each node in a linked list must contain at least _____ field(s).
- **(a)** Four
- **(b)** Three
- **(c)** One
- **(d)** Two

3. Which type of linked list stores the address of the header node in the address field of the last node?
- **(a)** Doubly linked list
- **(b)** Circular header linked list
- **(c)** Singly linked list
- **(d)** Header linked list

4. The situation in a linked list when START = NULL is:
- **(a)** Overflow
- **(b)** Underflow
- **(c)** Both
- **(d)** None of these

5. Linked lists can be implemented in what type of data structures?
   (a) Queues
   (b) Trees
   (c) Stacks
   (d) All of these

6. Which type of linked list contains a node to the next as well as the previous nodes?
   (a) Doubly linked list
   (b) Singly linked list
   (c) Circular linked list
   (d) Header linked list

7. The first node in the linked list is called _____.
   (a) End
   (b) Middle
   (c) Start
   (d) Begin

8. A linked list cannot grow and shrink during compile time.
   (a) False
   (b) It might grow
   (c) True
   (d) None of the above

9. What does NULL represent in the linked list?
   (a) Start of list
   (b) End of list
   (c) None of the above

CHAPTER 5

*Q*UEUES

5.1 Introduction

A queue is an important data structure which is widely used in many computer applications. A queue can be visualized with many examples from our day-to-day life with which we are already familiar. A very simple illustration of a queue is a line of people standing outside to enter a movie theatre. The first person standing in the line will enter the movie theatre first. Similarly, there are many daily life examples in which we can see the queue being implemented. Hence, we observe that whenever we talk about a queue, we see that that the element at the first position will be served first. Thus, a queue can be described as a FIFO (first in, first out) data structure; that is, the element which is inserted first will be the first one to be taken out. Now, let us discuss more about queues in detail.

5.2 Definition of a Queue

A queue is a linear collection of data elements in which the element inserted first will be the element taken out first (i.e., a queue is a FIFO data structure).A queue is an abstract data structure, somewhat similar to stacks. Unlike stacks, a queue is open on both ends. *A queue is a linear data structure in which the first element is inserted on one end called the REAR end (also called the tail end), and the deletion of the element takes place from the other end called the FRONT end (also called the head).*One end is always used to insert data and the other end is used to remove data.

Queues can be implemented by using arrays or linked lists. We will discuss the implementation of queues using arrays and linked lists in this section.

Practical Application

▪ A real-life example of a queue is people moving on an escalator. The people who got on the escalator first will be the first ones to step off of it.

▪ Another illustration of a queue is a line of people standing at the bus stop waiting for the bus. Therefore, the first person standing in the line will get into the bus first.

5.3   Implementation of a Queue

Queues can be represented by two data structures:

1. Representation of queues using arrays.

2. Representation of queues using linked lists.

Now, let us discuss both of them in detail.

5.3.1   Implementation of Queues Using Arrays

Queues can be easily implemented using arrays. Initially the front end (head) and the rear end (tail) of the queue point at the first position or location of the array. As we insert new elements into the queue, the rear keeps on incrementing, always pointing to the position where the next element will be inserted, while the front remains at the first position. The representation of a queue using an array is shown as follows:

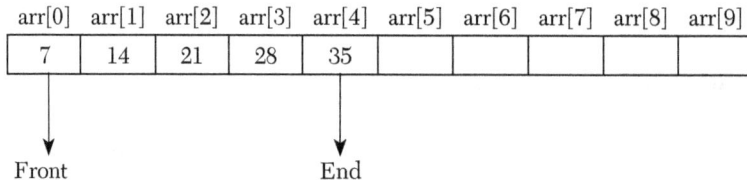

arr[0]	arr[1]	arr[2]	arr[3]	arr[4]	arr[5]	arr[6]	arr[7]	arr[8]	arr[9]
7	14	21	28	35					

Front   End

Figure 5.1. Array representation of a queue.

5.3.2   Implementation of Queues Using Linked Lists

We have already studied how a queue is implemented using an array. Now let us discuss the same using linked lists. We already know that in linked lists, dynamic memory allocation takes place; that is, the memory is allocated at runtime. But in the case of arrays, memory is allocated at the start of the program. This we have already discussed in the chapter about linked lists. If we are aware of the maximum size of the queue in advance, then implementation of a queue using arrays will be efficient. But if the size is not known in advance, then we will use the concept of a linked list, in which dynamic memory allocation takes place. As we all know a linked list has two parts, in which the first part contains the information of the node and the second part stores the address of the next element in the linked list. Similarly, we can also implement a linked queue. The START node in the linked list will become the FRONT node in a linked queue, and the end of the queue will be denoted by REAR. All insertion operations will be done at the rear end only. Similarly, all deletion operations will be done at the front end only.

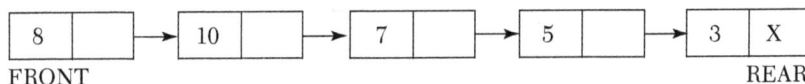

Figure 5.2. A linked queue.

5.3.2.1 Insertion in Linked Queues

Insertion is the process of adding new elements in the already existing queue. The new elements in the queue will always be inserted from the rear end. Initially, *we will check whether FRONT = NULL*. If the condition is true, then the queue is empty; otherwise, the new memory is allocated for the new node. We will understand it further with the help of an algorithm:

Algorithm for inserting a new element in a linked queue

```
Step 1: START
Step 2: Set NEW NODE.INFO = VAL
    IF FRONT = NULL
    Set FRONT = REAR = NEW NODE
    Set FRONT.NEXT = REAR.NEXT = NEW NODE
    ELSE
       Set REAR.NEXT = NEW NODE
       Set NEW NODE.NEXT = NULL
       Set REAR  = NEW NODE
    [End of If]
Step 3: EXIT
```

In the previous algorithm, first we are allocating the memory for the new node. Then we are initializing it with the information to be stored in it. Next, we are checking if the new node is the first node of the queue or not. If the new node is the first node of the queue, then we are storing NULL in the address part of the new node. In this case, the new node is tagged as FRONT as well as REAR. However, if the new node is not the first node of the queue, then in that case it is inserted at the REAR end of the queue.

For example: Consider a linked queue with five elements; a new element is to be inserted in the queue.

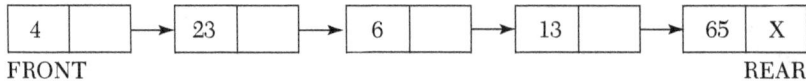

Figure 5.3. Linked queue before insertion.

After inserting the new element in the queue, the updated queue becomes as shown in the following figure:

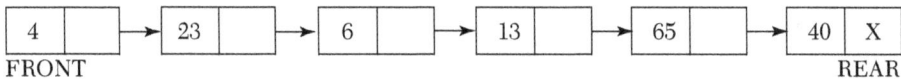

Figure 5.4. Linked queue after insertion.

5.3.2.2 Deletion in Linked Queues

Deletion is the process of removing elements from the already existing queue. The elements from the queue will always be deleted from the front end. Initially, we will check with the underflow condition, that is, whether FRONT = NULL. If the condition is true, then the queue is empty, which means we cannot delete any elements from it. Therefore, in that case an underflow error message is displayed on the screen. We will understand it further with the help of an algorithm:

Algorithm for deleting an element from a queue

```
Step 1: START
Step 2: IF FRONT = NULL
           Print UNDERFLOW ERROR
        [End of If]
Step 3: Set TEMP = FRONT
Step 4: Set FRONT = FRONT.NEXT
Step 5: EXIT
```

In the previous algorithm, we first check with the underflow condition, that is, whether the queue is empty or not. If the condition is true, then an underflow error message will be displayed; otherwise, we will use a node variable TEMP which will point to the FRONT. In the next step, FRONT is now pointing to the second node in the queue. Finally, the first node is deleted from the queue.

For example: Consider a linked queue with five elements; an element is to be deleted from the queue.

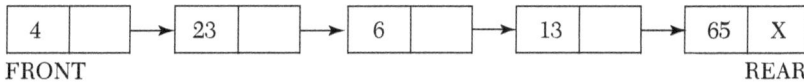
FRONT   REAR

Figure 5.5. Linked queue before deletion.

After deleting an element from the queue, the updated queue becomes as shown in the following figure:

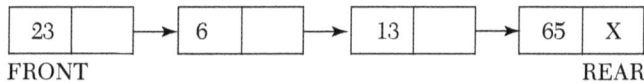
FRONT   REAR

Figure 5.6. Linked queue after deletion.

// Write a menu-driven program implementing a linked queue performing insertion and deletion operations.

```
public class Queue {
    private class Node {
        int data;
        Node next;
    }
    private Node front;
    private Node rear;
    public void insertion(int item) {
        Node = new Node();
        node.data = item;
        node.next = null;
        if (front == null) {
            front = rear = node;
        } else {
            rear.next = node;
            rear = node;
        }
    }
    public int deletion() throws Exception {
        if (front == null) {
            throw new Exception("Queue is Empty");
        }
        int temp = front.data;
        front = front.next;
        return temp;
    }
    public void display() throws Exception {
        if (front == null) {
            throw new Exception("Queue is Empty");
```

```java
        }
        Node temp = front;
        while (temp != null) {
            System.out.print(temp.data + "->");
            temp = temp.next;
        }
        System.out.print("NULL \n");
    }
}
//CLIENT CLASS
import java.util.Scanner;
public class QueueClient {
    public static void main(String[] args) {
        Scanner scn = new Scanner(System.in);
        Queue q = new Queue();
        boolean flag = true;
        try {
            while (flag) {
                System.out.println("\n***MENU***");
                System.out.println("1. Insertion");
                System.out.println("2. Deletion");
                System.out.println("3. Display");
                System.out.println("4. Exit");
                System.out.println("Enter your choice: ");
                int choice = scn.nextInt();
                int item = 0;
                switch (choice) {
                case 1:
                    System.out.println("Enter value of node: ");
                    item = scn.nextInt();
                    q.insertion(item);
                    System.out.println(item + " inserted
                                        successfully");
                    break;
                case 2:
                    System.out.println(q.deletion() +
                                    " deleted successfully");
                    break;
                case 3:
                    q.display();
                    break;
                case 4:
                    flag = false;
                    System.out.println("Terminated.....");
                    break;
```

```
            default:
                    System.out.println("Wrong choice");
                    break;
            }
        }
    } catch (Exception e) {
        System.out.println(e.getMessage());
    }
  }
}
```

The output of the program is shown as:

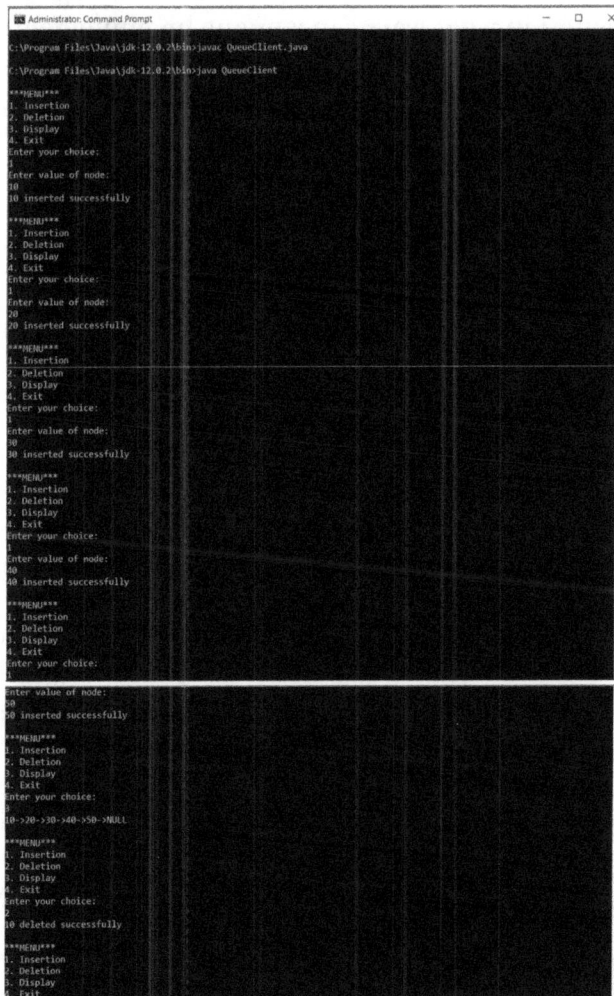

Frequently Asked Questions

1. Define queues; in what ways can a queue be implemented?

Ans: *A queue is a linear data structure in which the first element is inserted from one end called the REAR end (also called the tail end) and the deletion of the element takes place from the other end called the FRONT end (also called the head). Each type of queue can be implemented in two ways:*

- *Array Representation (Static Representation)*

- *Linked List Representation (Dynamic Representation)*

5.4 Operations on Queues

The two basic operations that can be performed on queues are as follows:

5.4.1 Insertion

Insertion is the process of adding new elements in the queue. However, before inserting any new element in the queue, we must always check for the overflow condition, which occurs when we try to insert an element in a queue which is already full. An overflow condition can be checked as follows: If REAR = MAX − 1, where MAX is the size of the queue. Hence, if the overflow condition is true, then an overflow message is displayed on the screen; otherwise, the element is inserted into the queue. Insertion is always done at the rear end. Insertion is also known as en-queue.

For example: Let us take a queue which has five elements in it. Suppose we want to insert another element, 50, in it; then REAR will be incremented by 1. Thus, a new element is inserted at the position pointed to by REAR. Now, let us see how insertion is done in the queue in the following figure:

arr[0]	arr[1]	arr[2]	arr[3]	arr[4]	arr[5]	arr[6]	arr[7]	arr[8]	arr[9]
7	14	21	28	35					

Front   End

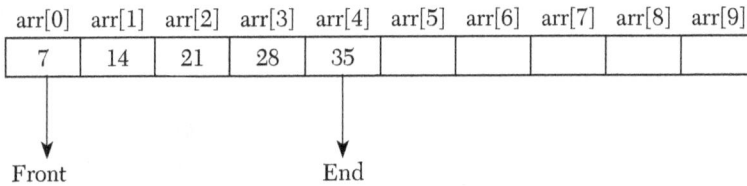

After inserting 50 in it, the new queue will be:

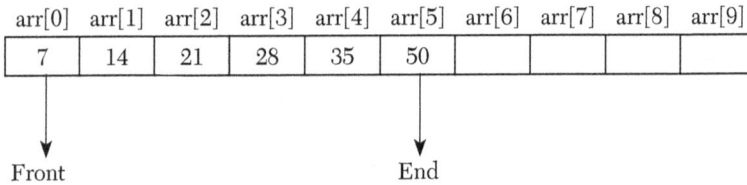

arr[0]	arr[1]	arr[2]	arr[3]	arr[4]	arr[5]	arr[6]	arr[7]	arr[8]	arr[9]
7	14	21	28	35	50				

Front   End

Figure 5.7. Queue after inserting a new element.

Algorithm for inserting a new element in a queue

```
Step 1: START
Step 2: IF REAR = MAX - 1
            Print OVERFLOW ERROR
        [End of If]
Step 3: IF FRONT = -1 && REAR = -1
            Set FRONT = 0
            Set REAR = 0
            ELSE
            REAR = REAR + 1
        [End of If]
Step 4: Set QUE[REAR] = ITEM
Step 5: EXIT
```

In the previous algorithm, first we check for the overflow condition. In Step 2, we are checking to see whether the queue is empty or not. If the queue is empty, then both FRONT and REAR are set to zero; otherwise, REAR is incremented to the next position in the queue. Finally, the new element is stored in the queue at the position pointed to by REAR.

5.4.2 Deletion

Deletion is the process of removing elements from the queue. However, before deleting any element from the queue, we must always check for the underflow condition, which occurs when we try to delete an element from a queue which is empty. An underflow condition can be checked as follows: If FRONT > REAR or FRONT = -1. Hence, if the underflow condition is true, then an underflow message is displayed on the screen; otherwise, the element is deleted from the queue. Deletion is always done at the front end. Deletion is also known as de-queue.

For example: Let us take a queue which has five elements in it. Suppose we want to delete an element,7, from a queue; then FRONT will be incremented by 1. Thus, the new element is deleted from the position pointed to by FRONT. Now, let us see how deletion is done in the queue in the following figure:

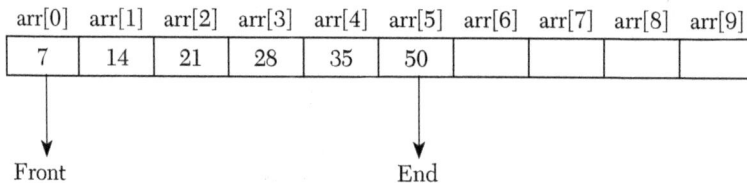

arr[0]	arr[1]	arr[2]	arr[3]	arr[4]	arr[5]	arr[6]	arr[7]	arr[8]	arr[9]
7	14	21	28	35	50				

Front   End

After deleting 7 from it, the new queue will be:

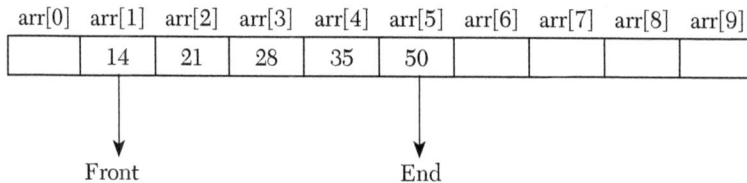

arr[0]	arr[1]	arr[2]	arr[3]	arr[4]	arr[5]	arr[6]	arr[7]	arr[8]	arr[9]
	14	21	28	35	50				

Front   End

Figure 5.8. Queue after deleting an element.

Algorithm for deleting an element from a queue

```
Step 1: START
Step 2: IF FRONT > REAR or FRONT = -1
            Print UNDERFLOW ERROR
        [End of If]
Step 3: Set ITEM = QUE[FRONT]
Step 4:Set FRONT = FRONT + 1
Step 5: EXIT
```

In the previous algorithm, first we check for the underflow condition, that is, whether the queue is empty or not. If the queue is empty, then no deletion takes place; otherwise, FRONT is incremented to the next position in the queue. Finally, the element is deleted from the queue.

// Write a menu-driven program for a linear queue, performing insertion and deletion operations.

```
public class SQueue {
    static int MAX = 50;
```

```java
    private int[] data;
    private int front;
    private int rear;
    public void insertion(int item) throws Exception {
        if (rear == MAX - 1) {
            throw new Exception("Queue is full");
        } else {
            data[rear++] = item;
        }
    }
    public int deletion() throws Exception {
        if (front == rear) {
            throw new Exception("Queue is Empty");
        }
        int temp = data[front];
        front++;
        return temp;
    }
    public void display() throws Exception {
        if (front == rear) {
            throw new Exception("Queue is Empty");
        }
        for (int i = front; i < rear; i++) {
            System.out.print(data[i] + " ");
        }
    }
}
//CLIENT CLASS
import java.util.Scanner;
public class SQClient {
    public static void main(String[] args) {
        Scanner scn = new Scanner(System.in);
        SQueue q = new SQueue();
        boolean flag = true;
        try {
            while (flag) {
                System.out.println("\n***MENU***");
                System.out.println("1. Insertion");
                System.out.println("2. Deletion");
                System.out.println("3. Display");
                System.out.println("4. Exit");
                System.out.println("Enter your choice: ");
                int choice = scn.nextInt();
                int item = 0;
```

```
            switch (choice) {
            case 1:
                System.out.println("Enter value of node: ");
                item = scn.nextInt();
                q.insertion(item);
                System.out.println(item + " inserted
                                        successfully");
                break;
            case 2:
                System.out.println(q.deletion() + " deleted
                                        successfully");
                break;
            case 3:
                q.display();
                break;
            case 4:
                flag = false;
                System.out.println("Terminated.....");
                break;
            default:
                System.out.println("Wrong choice");
                break;
            }
        }
    } catch (Exception e) {
        System.out.println(e.getMessage());
    }
    }
}
}
```

The output of the program is shown as:

5.5 Types of Queues

This section discusses various types of queues which include:

1. Circular Queue

2. Priority Queue

3. De-Queue (Double-ended queue)
Let us discuss all of them one by one in detail.

5.5.1 Circular Queue

A circular queue is a special type of queue which is implemented in a circular fashion rather than in a straight line. A circular queue is a linear data structure in which the operations are performed based on the FIFO (First In First Out) principle and the last position is connected to the first position to make a circle. It is also called a "ring buffer."

5.5.1.1 Limitation of Linear Queues

In linear queues, we studied how insertion and deletion takes place. We discussed that inserting a new element in the queue is only done at the rear end. Similarly, deleting an element from the queue is only done at the front end. Now let us consider a queue of ten elements given as follows:

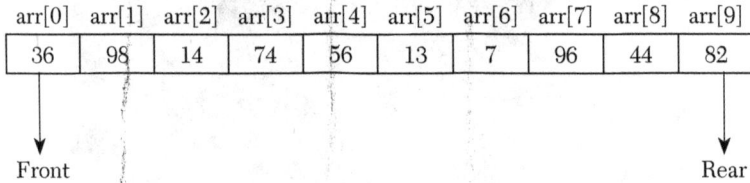

arr[0]	arr[1]	arr[2]	arr[3]	arr[4]	arr[5]	arr[6]	arr[7]	arr[8]	arr[9]
36	98	14	74	56	13	7	96	44	82

Front   Rear

The queue is now full, so we cannot insert any more elements in it. If we delete three elements from the queue, now the queue will be:

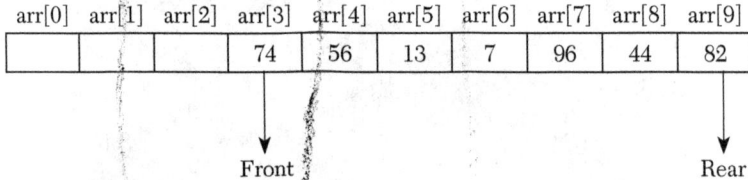

arr[0]	arr[1]	arr[2]	arr[3]	arr[4]	arr[5]	arr[6]	arr[7]	arr[8]	arr[9]
			74	56	13	7	96	44	82

Front   Rear

Thus, we can see that even after the deletion of three elements from the queue, the queue is still full, as REAR = MAX − 1. We still cannot insert any new elements in it as there is no space to store new elements. Therefore, this is a major drawback of the linear queue.

To overcome this problem, we can shift all the elements to the left so that the new elements can be inserted from the rear end, but shifting all the elements of the queue can be a very time-consuming procedure, as the practical queues are very large in size. Another solution to this problem is a circular queue. First, let us see how a circular queue looks, as in the following figure:

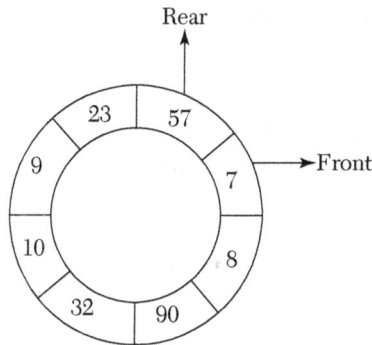

Figure 5.9. A circular queue.

In a circular queue, the elements are stored in a circular form such that the first element is next to the last element in the queue, as shown in the figure. A circular queue will be full when FRONT = 0 and REAR = MAX – 1 or FRONT = REAR + 1.In that case an overflow error message will be displayed on the screen. Similarly, a circular queue will be empty when both FRONT and REAR are equal to zero. In that case, an underflow error message will be displayed on the screen. Now, let us study both insertion and deletion operations on a circular queue.

Practical Application

A circular queue is used in operating systems for scheduling different processes.

Frequently Asked Questions

2. What is a circular queue? List the advantages of a circular queue over a simple queue.

Ans: *A circular queue is a particular kind of queue where new items are added to the rear end of the queue and items are read off from the front end of the queue, so there is a constant stream of data flowing in and out of the queue. A circular queue is also known as a "circular buffer." It is a structure that allows data to be passed from one process to another, making the most efficient use of memory. The only difference between a linear queue and circular queue is that in a linear queue when the rear points to the last position in the array, we cannot insert data even if we have deleted some elements. But in a circular queue we can insert elements as long as there is free space available. The main advantage of a circular queue as compared to a linear queue is that it avoids the wastage of space.*

5.5.1.2 Inserting an Element in a Circular Queue

While inserting a new element in the already existing queue, we will first check for the overflow condition, which occurs when we are trying to insert an element in the queue which is already full, as previously discussed. The position of the new element to be inserted can be calculated by using the following formula:

REAR = (REAR + 1) % MAX, *where MAX is equal to the size of the queue.*

For example: Let us consider a circular queue with three elements in it. Suppose we want to insert an element 56 in it. Let us see how insertion Is done in the circular queue.

Step 1: Initially the queue contains three elements. FRONT denotes the beginning of the circular queue, and REAR denotes the end of the circular queue.

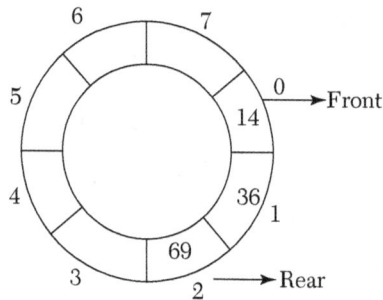

Figure 5.10. Initial circular queue without insertion.

Step 2: Now, the new element is to be inserted in the queue. Hence, REAR = REAR + 1;that is, REAR will be incremented by 1 so that it points to the next location in the queue.

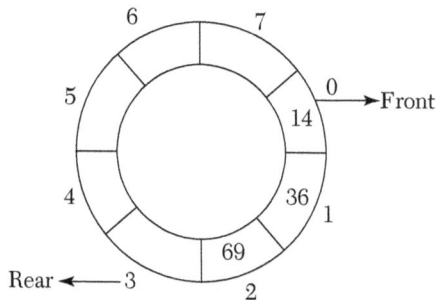

Figure 5.11. REAR is incremented by 1 so that it points to the next location.

Step 3: Finally, in this step the new element is inserted at the location pointed to by REAR. Hence, after insertion the queue is shown as in the following figure:

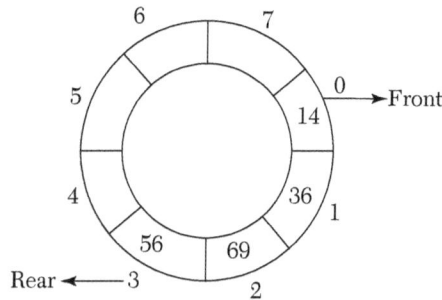

Figure 5.12. Final queue after inserting a new element.

Algorithm for inserting an element in a circular queue

Here QUEUE is an array with N elements. FRONT and REAR point to the front and rear elements of the queue. ITEM is the value to be inserted.

```
Step 1: START
Step 2: IF (FRONT = 0 && REAR = MAX - 1) OR (FRONT = REAR + 1)
            Print OVERFLOW ERROR
Step 3: ELSE
            IF (FRONT = -1)
            Set FRONT = 0
            Set REAR = 0
Step 4: ELSE
            IF (REAR = MAX - 1)
            Set REAR = 0
            ELSE
            REAR = REAR + 1
        [End of If]
        [End of If]
Step 5: Set CQUEUE[REAR] = ITEM
Step 6: EXIT
```

In the previous algorithm, first we check with the overflow condition. Second, we check if the queue is empty or not. If the queue is empty, then FRONT and REAR are set to zero. In Step 4, if REAR has reached its maximum capacity, then we set REAR = 0;otherwise, REAR is incremented by 1 so that it points to the next position where the new element is to be inserted. Finally, the new element is inserted in the queue.

5.5.1.3 Deleting an Element from a Circular Queue

While deleting an element from the already existing queue, we will first check for the underflow condition, which occurs when we are trying to delete an element from a queue which is empty. After deleting an element from the circular queue, the position of the FRONT end can be calculated by the following formula:

$$FRONT = (FRONT + 1) \% MAX, \text{ where MAX is equal to the size of the queue.}$$

For example: Let us consider a circular queue with seven elements in it. Suppose we want to delete an element 45 from it. Let us see how deletion is done in the circular queue.

Step 1: Initially the queue contains seven elements. FRONT denotes the beginning of the circular queue, and REAR denotes the end of the circular queue.

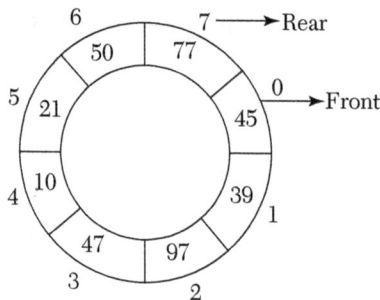

Figure 5.13. Initial circular queue without deletion.

Step 2: Now, the element is to be deleted from the queue. Hence, FRONT = FRONT + 1, that is, FRONT will be incremented by 1 so that it points to the next location in the queue. Also, the value is deleted from the queue. Thus, the queue after deletion is shown as follows:

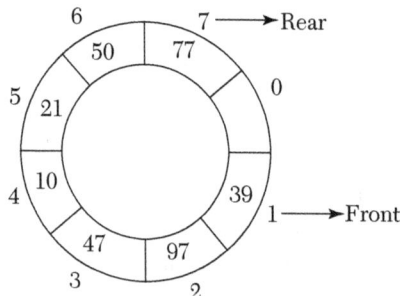

Figure 5.14. Final queue after deleting an element.

Algorithm for deleting an element from a circular queue

Here CQUEUE is an array with N elements. FRONT and REAR point to the front and rear elements of the queue. ITEM is the value to be deleted.

```
Step 1: START
Step 2: IF (FRONT = -1)
            Print UNDERFLOW ERROR
Step 3: ELSE
            Set ITEM = CQUEUE[FRONT]
Step 4:IF (FRONT = REAR)
            Set FRONT = -1
            Set REAR = -1
Step 5:ELSE IF (FRONT = MAX - 1)
            Set FRONT = 0
            ELSE
            FRONT = FRONT + 1
[End of If]
[End of If]
Step 6: EXIT
```

In the previous algorithm, we first check with the underflow condition. Second, we store the element to be deleted in ITEM. Third, we check to see if the queue is empty or not after deletion. Also, if FRONT has reached its maximum capacity, then we set FRONT = 0;otherwise, FRONT is incremented by 1 so that it points to the next position. Finally, the element is deleted from the queue.

// Write a menu-driven program for a linear circular queue performing insertion and deletion operations.

```java
public class CircularQueue {
    static int MAX = 50;
    private int[] data = new int[MAX];
    private int front;
    private int rear;
    public void insertion(int item) throws Exception {
        if ((front == 0 && rear == MAX - 1) ||
                                        (rear + 1 == front)) {
            throw new Exception("Queue is full");
        }
        if (rear == MAX - 1)
            rear = 0;
        else if (front == -1)
            front = 0;
```

```java
        else
            data[rear++] = item;
    }
    public int deletion() throws Exception {
        if (front == -1) {
            throw new Exception("Queue is Empty");
        }
        int temp = data[front];
        if (front == rear)
            front = rear = -1;
        else if (front == MAX - 1)
            front = 0;
        else
            front++;
        return temp;
    }
    public void display() throws Exception {
        if (front == -1) {
            throw new Exception("Queue is Empty");
        }
        if (rear < front) {
            for (int i = front; i <= MAX - 1; i++)
                System.out.print(data[i] + "->");
            for (int i = 0; i <= rear; i++)
                System.out.print(data[i] + "->");
        } else {
            for (int i = front; i < rear; i++) {
                System.out.print(data[i] + "->");
            }
        }
        System.out.println("NULL");
    }
}
//CLIENT CLASS
import java.util.Scanner;
public class CQClient {
    public static void main(String[] args) {
        Scanner scn = new Scanner(System.in);
        CircularQueue q = new CircularQueue();
        boolean flag = true;
        try {
            while (flag) {
                System.out.println("\n***MENU***");
```

```java
        System.out.println("1. Insertion");
        System.out.println("2. Deletion");
        System.out.println("3. Display");
        System.out.println("4. Exit");
        System.out.println("Enter your choice: ");
        int choice = scn.nextInt();
        int item = 0;
        switch (choice) {
        case 1:
            System.out.println("Enter value of node: ");
            item = scn.nextInt();
            q.insertion(item);
            System.out.println(item + " inserted
                                    successfully");
            break;
        case 2:
            System.out.println(q.deletion() + " deleted
                                    successfully");
            break;
        case 3:
            q.display();
            break;
        case 4:
            flag = false;
            System.out.println("Terminated.....");
            break;
        default:
            System.out.println("Wrong choice");
            break;
        }
    }
} catch (Exception e) {
    System.out.println(e.getMessage());
}
    }
}
```

The output of the program is shown as:

```
Administrator: Command Prompt                                    -   □   ×

C:\Program Files\Java\jdk-12.0.2\bin>javac CQClient.java

C:\Program Files\Java\jdk-12.0.2\bin>java CQClient

***MENU***
1. Insertion
2. Deletion
3. Display
4. Exit
Enter your choice:
1
Enter value of node:
25
25 inserted successfully

***MENU***
1. Insertion
2. Deletion
3. Display
4. Exit
Enter your choice:
1
Enter value of node:
36
36 inserted successfully

***MENU***
1. Insertion
2. Deletion
3. Display
4. Exit
Enter your choice:
1
Enter value of node:
49
49 inserted successfully

***MENU***
1. Insertion
2. Deletion
3. Display
4. Exit
Enter your choice:
1
Enter value of node:
58
58 inserted successfully

***MENU***
1. Insertion
2. Deletion
3. Display
4. Exit
Enter your choice:
1
Enter value of node:
67
67 inserted successfully

***MENU***
1. Insertion
2. Deletion
3. Display
4. Exit
Enter your choice:
3
25->36->49->58->67->NULL

***MENU***
1. Insertion
2. Deletion
3. Display
4. Exit
Enter your choice:
2
25 deleted successfully

***MENU***
1. Insertion
2. Deletion
3. Display
4. Exit
Enter your choice:
2
36 deleted successfully

***MENU***
1. Insertion
2. Deletion
3. Display
4. Exit
Enter your choice:
3
49->58->67->NULL

***MENU***
1. Insertion
2. Deletion
3. Display
4. Exit
Enter your choice:
4
Terminated.....

C:\Program Files\Java\jdk-12.0.2\bin>
```

5.5.2   Priority Queue

A priority queue is another variant of a queue, in which elements are processed on the basis of assigned priority. Each element in a priority queue is assigned a special value called the priority of the element. The elements in the priority queue are processed based on the following rules:

1. An element with higher priority is processed first, and then the element with lower priority is processed.

2. If the two elements have the same priority, then the elements are processed on the First Come First Served basis. The priority of the element is selected by its value called the implicit priority, and the priority number given with each element is called the explicit priority.

A priority queue is like a modified queue or stack data structure, but where additionally each element has a "priority" associated with it. In a priority queue, insertion and deletion operations are also done according to the assigned priority. If we want to delete an element from the priority queue, then the element with the highest priority is processed first and is deleted. The case is the same with insertion. The priority given to the elements in the queue is based on several factors. Priority queues are commonly used in operating systems for executing higher priority processes first. The priority assigned to these processes may be based on the time taken by the CPU to execute these processes completely.

Practical Application

In an operating system, if there are four processes to be executed where the first process needs 3ns to complete, the second process needs 5 ns to complete, the third process needs 9 ns to complete, and the fourth needs 8 ns to complete, then the first process will be given the highest priority and will be the first to be executed among all the processes.

Now the priority queues are further divided into two types which are:

1. Ascending Priority Queue: In this type of priority queue, elements can be inserted in any order, but at the time of deletion of elements from the queue, the smallest element is searched and deleted first.

2. Descending Priority Queue: In this type of priority queue, elements can be inserted in any order. But at the time of deletion of elements from the queue, the largest element is searched and deleted first. For example: Operating systems, Routing.

Frequently Asked Questions

3. Define Priority Queue.

Ans: *A priority queue is a collection of elements such that each element has been assigned a priority and such that the order in which elements are deleted and processed comes from the following rules:*

(a) *An element of higher priority is processed before any element of lower priority.*

(b) *Two elements with same priority are processed according to the order in which they were added to the queue.*

The array elements in a priority queue can have the following structure:

```
private class Node {
    int priority;
    int data;
    Node next;
    }
```

5.5.2.1 Implementation of a Priority Queue

A priority queue can be implemented in two ways:

1. Array Representation of a Priority Queue

2. Linked Representation of a Priority Queue

Let us now discuss both these implementations in detail.

1. Implementation of a priority queue using arrays

While implementing a priority queue using arrays, the following points must be considered:

■ Maintain a separate queue for each level of priority or priority number.

■ Each queue will appear in its own circular array and must have its own pairs of nodes, that is, FRONT AND REAR.

■ If each queue is allocated the same amount of memory, then a 2D array can be used instead of a linear array.

For example: FRONT [K] and REAR [K] are the nodes containing the front and rear values of row "K" of the queue, where K is the priority number. If we want to insert an element with priority K, then we will add the element at the REAR end of row K; K is the row as well as the priority number of that element. If we add F with priority number 4, then the queue will be given as shown in the following:

FRONT	REAR
2	2
1	3
0	0
5	1
4	4

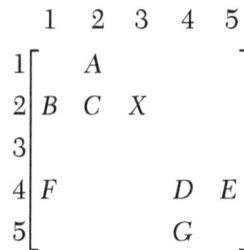

$$
\begin{array}{c}
\quad\ 1\ \ 2\ \ 3\ \ 4\ \ 5 \\
\begin{array}{c}1\\2\\3\\4\\5\end{array}
\left[\begin{array}{ccccc}
& A & & & \\
B & C & X & & \\
& & & & \\
F & & & D & E \\
& & & G &
\end{array}\right]
\end{array}
$$

Figure 5.15. Priority queue after inserting a new element.

2. Implementation of a priority queue using linked lists

A priority queue can be implemented using a linked list. While implementing the priority queue using a linked list, every node will have three parts:

(a) Information part

(b) Priority number of the element

(c) Address of the next element

An element with higher priority will precede the element having lower priority. Also, priority number and priority are opposite to each other; that is, an element having a lower priority number means it has higher priority. For example, if there are two elements X and Y with priority numbers 2 and 7 respectively, then X will be processed first because it has higher priority.

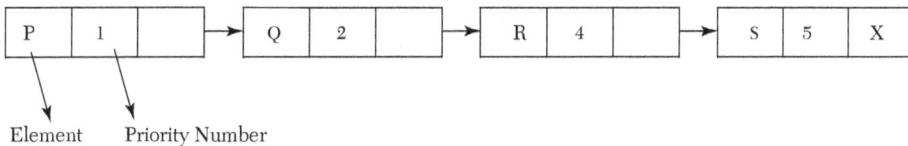

Element   Priority Number

Figure 5.16. A linked priority queue.

5.5.2.2 Insertion in a Linked Priority Queue

While inserting a new element in a linked priority queue, first we will traverse the entire queue until we find a node which has a lower priority than the new element. Thus, the new element is inserted before the element with the lower priority. Also, if there is an element in the queue which has same priority as that of the new element, then in that case the new element is inserted after that element.

For example: Consider a priority queue with four elements given as follows:

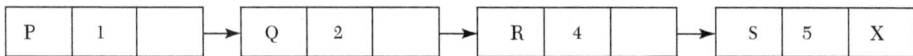

Figure 5.17. Linked priority queue before insertion.

Now, a new element with information A and priority number 3 is to be inserted; hence, the element will be inserted before R that has priority number 4, which is lower than that of the new element. The priority queue after inserting a new element is shown as follows:

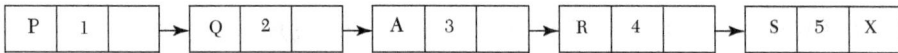

Figure 5.18. Linked priority queue after inserting a new element.

5.5.2.3 Deletion in a Linked Priority Queue

Deleting an element from a linked priority queue is a very simple process. In that case, the first node from the priority queue is deleted and the information of that node is processed first.

For example: Consider a priority queue with five elements given as follows:

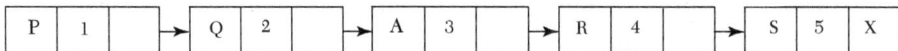

Figure 5.19. Linked priority queue before deletion.

Now, the first node from the queue is deleted. So, the priority queue after deletion is shown as follows:

Figure 5.20. Linked priority queue after deleting the first node.

```java
// Write a menu-driven program for a priority queue performing insertion
   and deletion operations.
public class PriorityQueue {
    private class Node {
```

```
        int priority;
        int data;
            Node next;
    }
    private Node front;
    public void insertion(int item, int prior) {
        Node = new Node();
        node.data = item;
        node.priority = prior;
        node.next = null;
        if (front == null || prior <= front.priority) {
            node.next = front;
            front = node;
        } else {
            Node temp = front;
            while (temp.next != null && (temp.next).priority
                                                <= prior) {
                temp = temp.next;
            }
            node.next = temp.next;
            temp.next = node;
        }
    }
    public int deletion() throws Exception {
        if (front == null) {
            throw new Exception("Queue is Empty");
        }
        int temp = front.data;
        front = front.next;
        return temp;
    }
    public void display() throws Exception {
        if (front == null) {
            throw new Exception("Queue is Empty");
        }
        Node temp = front;
        System.out.println("Priority \t\t Value");
        while (temp != null) {
            System.out.println(temp.priority + " \t\t\t " +
                                    temp.data);
            temp = temp.next;
        }
    }
}
```

```java
//CLIENT CLASS
import java.util.Scanner;
public class PQClient {
    public static void main(String[] args) {
        Scanner scn = new Scanner(System.in);
        PriorityQueue q = new PriorityQueue();
        boolean flag = true;
        try {

    while (flag) {
                System.out.println("\n***MENU***");
                System.out.println("1. Insertion");
                System.out.println("2. Deletion");
                System.out.println("3. Display");
                System.out.println("4. Exit");
                System.out.println("Enter your choice: ");
                int choice = scn.nextInt();
                int item = 0, prior = 0;
                switch (choice) {
                case 1:
                    System.out.println("Enter value of node: ");
                    item = scn.nextInt();
                  System.out.println("Enter priority of node: ");
                    prior = scn.nextInt();
                    q.insertion(item, prior);
                    System.out.println(item + " inserted
                                            successfully");
                    break;
                case 2:
                    System.out.println(q.deletion() + " deleted
                                            successfully");
                    break;
                case 3:
                    q.display();
                    break;
                case 4:
                    flag = false;
                    System.out.println("Terminated.....");
                    break;
                default:
                    System.out.println("Wrong choice");
                    break;
                }
```

```
        }
    } catch (Exception e) {
        System.out.println(e.getMessage());
    }
  }
}
```

The output of the program is shown as:

5.5.3 De-Queues (Double-Ended Queues)

A double-ended queue (de-queue, pronounced "deck") is a special type of data structure in which insertion and deletion of elements is done at either end, that is, either at the front end or at the rear end of the queue. It is often called ahead-tail linked list, because elements are added or removed from either the head (front) end or tail (end). De-queues are implemented using circular arrays in the computer's memory. The LEFT and RIGHT nodes are maintained in the de-queue, which point to either end of the queue.

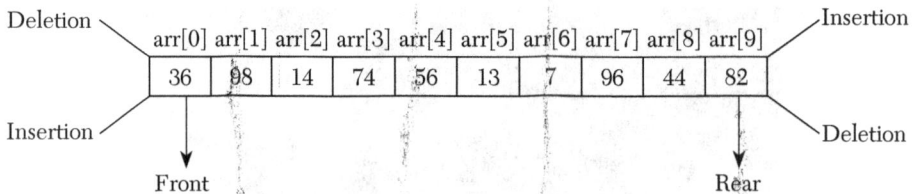

Figure 5.21. A double-ended queue.

Practical Application

A real-life example of a de-queue is that in a train station, the entry and exit of passengers can take place from both sides.

There are two types of double-ended queues, which include:

1. **Input Restricted De-Queue:** In this, the deletion operation can be performed at both ends (i.e., both front and rear end) while the insertion operation can be performed only at one end(i.e., rear end).

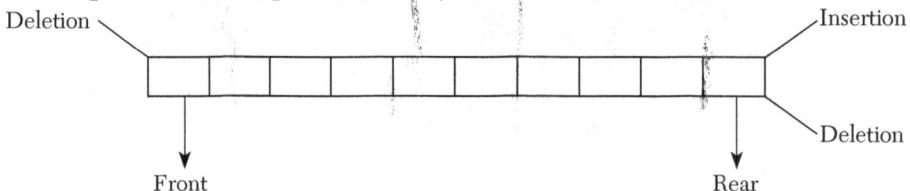

Figure 5.22. An input restricted double-ended queue.

2. Output Restricted De-Queue: In this, the insertion operation can be performed at both ends while the deletion operation can be performed only at one end (i.e., front end).

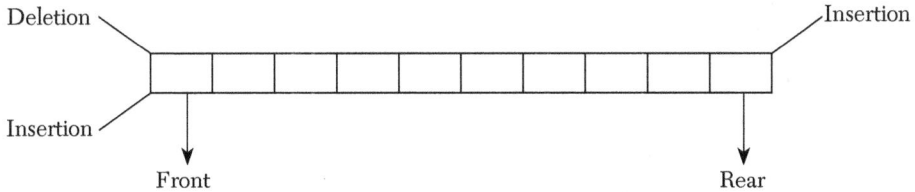

Figure 5.23. An output restricted double-ended queue.

// Write a menu-driven program for a double-ended queue performing insertion and deletion operations.

```
public class DQueue {
    static int MAX = 10;
    private int[] data = new int[MAX];
    private int front=-1;
    private int rear=-1;
    public void insertion_at_beginning(int item) throws Exception {
        if (rear == MAX - 1) {
            throw new Exception("Queue is full");
        }
        if (front == -1) {
            front = rear = 0;
            data[front] = item;
        } else {
            for(int i = rear;i>=front;i--)
                    data[i]=data[i+1];
            data[front]=item;
            display();
        }
    }
    public void insertion_at_end(int item) throws Exception {
        if (rear == MAX - 1) {
            throw new Exception("Queue is full");
        }
        data[++rear] = item;
    }
    public int deletion_from_beginning() throws Exception {
        if (front == rear) {
            throw new Exception("Queue is Empty");
        }
        int temp = data[front];
```

```
            front++;
            return temp;
    }
    public int deletion_from_end() throws Exception {
        if (front == rear) {

    throw new Exception("Queue is Empty");
        }
        int temp = data[rear];
        rear--;
        return temp;
    }
  public void display() throws Exception {
        if (front == -1) {
            throw new Exception("Queue is Empty");
        }
        for (int i = front; i < rear; i++) {
            System.out.print(data[i] + "->");
        }
        System.out.println("NULL");
    }
}
//CLIENT CLASS
import java.util.Scanner;
public class DQClient {
    public static void main(String[] args) {
        Scanner scn = new Scanner(System.in);
        DQueue q = new DQueue();
        boolean flag = true;
        try {
            while (flag) {
                System.out.println("\n***MENU***");
                System.out.println("1. Insertion at beginning");
                System.out.println("2. Insertion at end");
                System.out.println("3. Deletion at beginning");
                System.out.println("4. Deletion at end");
                System.out.println("5. Display");
                System.out.println("6. Exit");
                System.out.println("Enter your choice: ");
                int choice = scn.nextInt();
                int item = 0;
                switch (choice) {
```

```
            case 1:
                System.out.println("Enter value of node: ");
                item = scn.nextInt();
                q.insertion_at_beginning(item);
                System.out.println(item + " inserted
                                        successfully");
                break;
            case 2:
                System.out.println("Enter value of node: ");
                item = scn.nextInt();
                q.insertion_at_end(item);
                System.out.println(item + " inserted
                                        successfully");
    break;
            case 3:
                    System.out.println(q.deletion_from_begin-
ning() + " deleted successfully");
    break;
            case 4:
                System.out.println(q.deletion_from_end() +
                                " deleted successfully");
                break;
            case 5:
                q.display();
                break;
            case 6:
                flag = false;
                System.out.println("Terminated.....");
                break;
            default:
                System.out.println("Wrong choice");
                break;
            }
        }
    } catch (Exception e) {

        System.out.println(e.getMessage());
    }
  }
}
```

The output of the program is shown as:

```
Administrator: Command Prompt                                          -    □    ×

C:\Program Files\Java\jdk-12.0.2\bin>javac DQClient.java

C:\Program Files\Java\jdk-12.0.2\bin>java DQClient

***MENU***
1. Insertion at beginning
2. Insertion at end
3. Deletion at beginning
4. Deletion at end
5. Display
6. Exit
Enter your choice:
1
Enter value of node:
10
10 inserted successfully

***MENU***
1. Insertion at beginning
2. Insertion at end
3. Deletion at beginning
4. Deletion at end
5. Display
6. Exit
Enter your choice:
2
Enter value of node:
20
```

```
20 inserted successfully

***MENU***
1. Insertion at beginning
2. Insertion at end
3. Deletion at beginning
4. Deletion at end
5. Display
6. Exit
Enter your choice:
2
Enter value of node:
30
30 inserted successfully

***MENU***
1. Insertion at beginning
2. Insertion at end
3. Deletion at beginning
4. Deletion at end
5. Display
6. Exit
Enter your choice:
2
Enter value of node:
40
40 inserted successfully
```

```
***MENU***
1. Insertion at beginning
2. Insertion at end
3. Deletion at beginning
4. Deletion at end
5. Display
6. Exit
Enter your choice:
5
10->20->30->40->NULL

***MENU***
1. Insertion at beginning
2. Insertion at end
3. Deletion at beginning
4. Deletion at end
5. Display
6. Exit
Enter your choice:
3
10 deleted successfully

***MENU***
1. Insertion at beginning
2. Insertion at end
3. Deletion at beginning
```

```
4. Deletion at end
5. Display
6. Exit
Enter your choice:
4
40 deleted successfully

***MENU***
1. Insertion at beginning
2. Insertion at end
3. Deletion at beginning
4. Deletion at end
5. Display
6. Exit
Enter your choice:
5
20->30->NULL

***MENU***
1. Insertion at beginning
2. Insertion at end
3. Deletion at beginning
4. Deletion at end
5. Display
6. Exit
Enter your choice:
6
Terminated.....

C:\Program Files\Java\jdk-12.0.2\bin>
```

5.6   Applications of Queues

- In real life, call center phone systems use queues to hold people calling them in an order until a service representative is free.

- The handling of interrupts in real-time systems uses the concept of queues. The interrupts are handled in the same order as they arrive, that is, First Come First Served.

- The round-robin technique for processor scheduling is implemented using queues.

- Queues are often used as buffers on portable CD players, MP3 players, and in iPod playlists.

5.7   Summary

- A queue is a linear collection of data elements in which the element inserted first will be the element taken out first (i.e., a queue is a FIFO data structure).

- A queue is a linear data structure in which the first element is inserted from one end called the REAR end, and the deletion of the element takes place from the other end called the FRONT end.

- The implementation of queues can be done in two ways, which are implementation through arrays and implementation through linked lists.

- Insertion and deletion are the two basic operations that are performed on the queues.

- A circular queue is a linear data structure in which the operations are performed based on a FIFO (First In First Out) principle and the first index comes after the last index.

- A priority queue is a queue in which elements are processed on the basis of assigned priority. Each element in a priority queue is assigned a special value called the priority of the element.

- When a priority queue is implemented using linked lists, then every node of the list will have three parts, that is, a data part, the priority number of the element, and the address of the next element.

- A double-ended queue is a special type of data structure in which insertion and deletion of elements is done at either end, that is, either at the front end or at the rear end of the queue.

- An input restricted de-queue is a queue in which deletion can be done at both ends, but insertion is done only at the rear end.

- An output restricted de-queue is a queue in which insertion can be done at both ends, but deletion is done only at the front end.

5.8 Exercises

5.8.1 Theory Questions

1. What is a linear queue? Give its real-life example.

2. What is a circular queue and how it is different from a linear queue?

3. Define priority queues.

4. Discuss various operations which can be performed on the queues.

5. Define queues and in what ways a queue can be implemented. What do you understand about double-ended queues? Discuss the different types of de-queues in detail.

6. Give some of the applications of queues.

7. Why are queues known as First-In-First-Out structures?

8. Explain the concept of a linked queue and also discuss how insertion and deletion take place in it.

5.8.2 Programming Questions

1. Write a program to create a linear queue containing nine elements.

2. Write an algorithm to implement a priority queue.

3. Write a code for insertion and deletion in a queue.

4. Give an algorithm for insertion of an element in a circular queue. Write a program to implement a queue which allows insertion and deletion at both ends.

5. Write an algorithm that reverses the elements of a queue.

6. Write an algorithm for insertion and deletion in a queue. Write the functions for insertion and deletion operations performed in a de-queue. Consider all possible cases.

7. Write a code for deleting an element from a circular queue.

8. Write a program to implement a priority queue using a linked list.

5.8.3 Multiple Choice Questions

1. New elements in the queue are always inserted from:

   (a) Front end

   (b) Middle

   (c) Rear end

   (d) Both (a) and (c)

2. A queue is a _____ data structure.

   (a) FIFO

   (b) LIFO

   (c) FILO

   (d) LILO

3. The overflow condition in the circular queue exists when:

   (a) Front = MAX – 1 and Rear = 0

   (b) Front = 0 and Rear = MAX – 1

   (c) Front = 0 and Rear = 0

   (d) Front = MAX – 1 and Rear = MAX – 1

4. If the elements P, Q, R, and S are placed in a queue and are deleted one by one, in what order will they be deleted?

   (a) PQRS

   (b) SRQP

   (c) PRQS

   (d) SRQP

5. A data structure in which elements are inserted or deleted from the front as well as from the rear end is:

   (a) Linear queue

   (b) De-queue

   (c) Priority Queue

   (d) Circular Queue

6. A line outside a movie theater represents a _____.

   (a) Linked List

   (b) Array

   (c) Queue

   (d) Stack

7. In a queue, deletion is always done at the _____.

   (a) Top end

   (b) Back end

   (c) Front end

   (d) Rear end

8. In a priority queue, two elements with the same priority are processed on a FCFS basis.

   (a) False

   (b) True

9. The function that inserts the elements in a queue is called _____.

   (a) Push

   (b) En-queue

   (c) Pop

   (d) De-queue

10. Which of the implementation of queues is better when the size of the queue is not known in advance?

   (a) Linked List Representation

   (b) Array Representation

   (c) Both

   (d) None of the above

CHAPTER 6

SEARCHING AND SORTING

6.1 Introduction to Searching

As we all know, computer systems are often used to store large numbers. We require some search mechanism to retrieve a specific record from the large amounts of data stored in our computer system. Searching means to find whether a particular data item exists in an array/list or not. *The process of finding a particular value in a list or an array is called searching.* If that particular value is present in the array, then the search is said to be successful and the location of that particular value is returned by the searching process. However, if the value does not exist, then searching is said to be unsuccessful. There are many different searching algorithms, but three of the popular searching techniques are as follows:

- Linear Search or Sequential Search

- Binary Search

- Interpolation Search

Here, we will discuss all these methods in detail.

6.2 Linear Search or Sequential Search

A linear search is also called a sequential search. This is a very simple technique used to search for a particular value in an array. *A linear search works by comparing the value of the key being searched for with every element of the array in a linear sequence until a match is found.* A search will be unsuccessful if all the data elements are readand the desired element is not found. The following are some important points:

- It is the simplest way to search an element in a list.
- It searches the data element sequentially, no matter whether the array is sorted or unsorted.

For example: Let us take an array of ten elements, which is declared as follows:

```
int array[10] = {87, 25, 14, 39, 74, 1, 99, 12, 30, 67};
```

and the value to be searched for in the array is VALUE = 74, and then search to find whether 74 exists in the array or not. If the value is present, then its position is returned. Here the position of VAL = 74 is POS = 4 (index starting from zero), which has been shown by the following figures:

Pass 1: 87 is compared with 74. Since 87 is not equal to 74, we will move to the next pass.

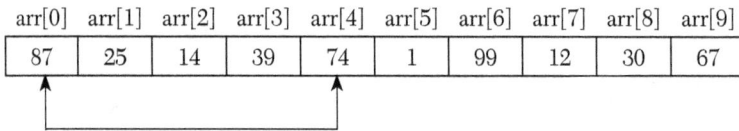

arr[0]	arr[1]	arr[2]	arr[3]	arr[4]	arr[5]	arr[6]	arr[7]	arr[8]	arr[9]
87	25	14	39	74	1	99	12	30	67

Pass 2: 25 is compared with 74. Since 25is not equal to 74, we will move to the next pass.

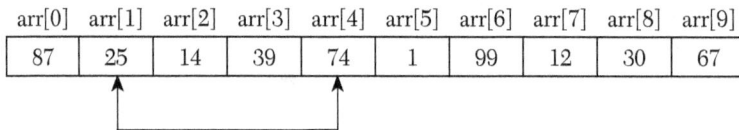

arr[0]	arr[1]	arr[2]	arr[3]	arr[4]	arr[5]	arr[6]	arr[7]	arr[8]	arr[9]
87	25	14	39	74	1	99	12	30	67

Pass 3: 14 is compared with 74. Since 14 is not equal to 74, we will move to the next pass.

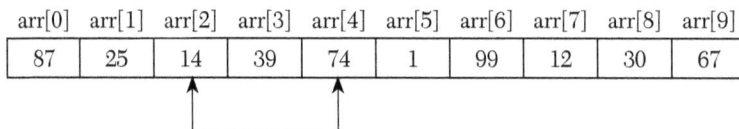

arr[0]	arr[1]	arr[2]	arr[3]	arr[4]	arr[5]	arr[6]	arr[7]	arr[8]	arr[9]
87	25	14	39	74	1	99	12	30	67

Pass 4: 39 is compared with 74. Since 39 is not equal to 74, we will move to the next pass.

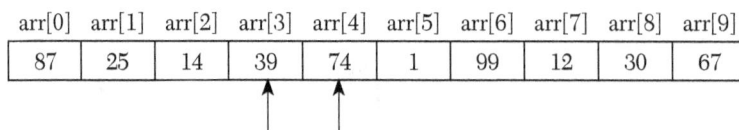

arr[0]	arr[1]	arr[2]	arr[3]	arr[4]	arr[5]	arr[6]	arr[7]	arr[8]	arr[9]
87	25	14	39	74	1	99	12	30	67

Pass 5: 74 is compared with 74. Since 74 is equal to 74, we will return the position on which 74 is present, which in this case is 4.

arr[0]	arr[1]	arr[2]	arr[3]	arr[4]	arr[5]	arr[6]	arr[7]	arr[8]	arr[9]
87	25	14	39	74	1	99	12	30	67

74 is found at POS = 4

Figure 6.1. Working of a linear search.

In this way, a linear search is used to search for a particular value in the array. Now let us understand it further with the help of an algorithm.

Practical Application

A simple and a real-life example of a linear search is that a person is searching for another person's contact number in a telephone directory. So, if the person does not know the exact name of that person but knows that the name starts with A, then he/she will start searching from the beginning of the telephone directory.

Algorithm for a Linear Search

Let *ARR* be an array of n elements, ARR[1], ARR[2], ARR[3], . . . ARR[n] such that VAL is the element to be searched. Then the algorithm will find the position POS of the VAL in the array ARR.

```
Step 1: START
Step 2: Set I = 0, POS = -1
Step 3: Repeat while I<N
            IF (ARR[I] = VAL)
            POS = I
        PRINT POS
            Go to Step 5
        [End of IF]
        [End of Loop]
Step 4: IF (POS = -1)
        PRINT "VALUE NOT FOUND, SEARCH UNSUCCESSFUL"
        [End of IF]
Step 5: EXIT
```

In Step 2 of the algorithm, we are initializing the values of I and POS. In Step 3, a while loop is executed in which a check is made to see whether a match is found between the current array element and VAL. If the match is found, then the position of that element is printed. In the last step, if all the elements have been compared and there is no match found, the search will be unsuccessful; that is, the value is not present in the array.

Complexity of a Linear Search Algorithm

*The execution time of a linear search is O(n),*where n is the number of elements in the array. The algorithm is called a linear search because its complexity can be expressed as a linear function, which is that the number of comparisons to find the target item increases linearly with the size of the data. The best case of a linear search is when the data element to be searched for is equal to the first element of the array. Obviously, the worst case will happen when the data element to be searched for is equal to the last element in the array. However, in both the cases n comparisons have to be made.

6.2.1 Drawbacks of a Linear Search

- It is a very time-consuming process, as it works sequentially.

- It can be applied only to a small amount of data.

- It is a very slow process, as almost every data element is accessed, especially when the data element is located near the end.

//Write a program to search for an element in an array using a linear search technique.

```
import java.util.Scanner;
public class LinearSearchDemo {
    static Scanner scn = new Scanner(System.in);
    public static void main(String[] args) {
        int[] array = takeInput();
        System.out.print("\nEnter value to search:");
        int val = scn.nextInt();
        int pos = linear_search(array, val);
        if (pos < 0) {
            System.out.print("\n" + val + " not found");
        } else {
            System.out.print("\n" + val + " fount at position " +
                            (pos + 1));
        }
    }
    public static int linear_search(int[] array, int value) {
        for (int i = 0; i < array.length; i++) {
            if (array[i] == value) {
                return i;
```

```
            }
        }
        return -1;
    }
    public static int[] takeInput() {
        System.out.print("\nEnter no of elements of array:");
        int size = scn.nextInt();
        int[] arr = new int[size];
        System.out.print("\nEnter the elements of array:");
        System.out.println();
        for (int i = 0; i < arr.length; i++) {
            System.out.print("Enter element "+(i+1)+": ");
            arr[i] = scn.nextInt();
        }
        return arr;
    }
}
```

The output of the program is shown as:

Frequently Asked Questions

Explain how a linear search technique is used to search for an element.

Ans: *Suppose that ARR is an array having N elements. ITEM is the value to be searched. Then we have the following cases:*

Case 1: Unsorted List: The ITEM is compared with every element of the array. If the element is found, then no further comparison is required. If all the elements are compared and checked, then the ITEM is not found.

Case 2: Sorted List: The ITEM is greater than the first element and smaller than the last element of the list, so searching is performed by comparing each element in the list with ITEM; otherwise, ITEM is reported as "Not Found."

6.3   Binary Search

A binary search is an extremely efficient searching algorithm when it is compared to a linear search. *A binary search works only when the array/list is already sorted.* In a binary search, we first compare the value VAL with the data element in the middle position of the array. If the match is found, then the position POS of that element is returned; otherwise, if the value is less than that of the middle element, then we begin our search in the lower half of the array and vice versa. So, we repeat this process on the lower and upper half of the array.

6.3.1   Binary Search Algorithm

Let us now understand how this binary search algorithm works in an array.

1. Find the middle element of the array, that is, n/2 is the middle element of the array containing n elements.

2. Now, compare the middle element of the array with the data element to be searched.

    (a) If the middle element is the desired element, then the search is successful.

    (b) If the data element to be searched for is less than the middle element of the array, then search only the lower half of the array, that is, those elements which are on the left side of the middle element.

    (c) If the data element to be searched for is greater than the middle element of the array, then search only the upper half of the array, that is, those elements which are on the right side of the middle element.

Repeat these steps until a match is found.

Practical Application

A real-life application of a binary search is that when we search for a particular word in a dictionary, we first open the dictionary somewhere in the middle. Now we will compare the desired word with the first word on that page. If the desired word comes after the first word on an open page, then we will look in the second half of the dictionary; otherwise, we will look in the first half. Now, we will again open a page in the second half and compare the desired word with the first word on that page, and the same process is repeated until we have found the desired word.

Algorithm for a Binary Search

Binary_Search(ARR, Lower_bound, Upper_bound, VAL)

```
Step 1: START
Step 2: Set BEG = lower_bound, END = upper_bound, POS = -1
Step 3: Repeat Steps 4 & 5 while BEG <= END
Step 4: Set MID = (BEG+END)/2
Step 5: IF (ARR[MID] = VAL)
            POS = MID
            PRINT POS
            Go to Step 7
            ELSE IF (ARR[MID] > VAL)
            Set END = MID - 1
            ELSE
            Set BEG = MID + 1
            [End of If]
            [End of Loop]
Step 6: IF (POS = -1)
                PRINT "VALUE NOT FOUND, SEARCH UNSUCCESSFUL"
            [End of IF]
Step 7: EXIT
```

In Step 2 of the algorithm, we are initializing the values of BEG, END, and POS. In Step 3 a while loop is executed. In Step 3, the value of MID is calculated. In Step 4 we will check if the value to be searched for is equal to the array value at MID. If the match is found, then the position of that element is printed. If the match is not found and the value to be searched for is less than that of the array value at MID, then the END is modified; otherwise, if the value to be searched for is greater than that of the array value at MID, then the BEG is modified. In the last step, if all the elements have been compared and there is no match found, the search has been unsuccessful; that is, the value is not present in the array.

For Example:

Let us now consider an example to search for a particular value in a sorted array.

Consider an array of ten elements which is declared as:

```
int array[10] = {0, 10, 20, 30, 40, 50, 60, 80, 90, 100};
```

and the value to be searched for is VAL = 20. Then the algorithm will proceed as follows:

Solution:

Pass 1:

$$\text{BEG} = 0, \text{END} = 10$$

$$\text{MID} = (\text{BEG} + \text{END})/2$$

$$= (0 + 10)/2 = 5$$

Now, VAL = 20 and ARR[MID] = ARR[5] = 50

arr[0]	arr[1]	arr[2]	arr[3]	arr[4]	arr[5]	arr[6]	arr[7]	arr[8]	arr[9]
0	10	20	30	40	50	60	70	80	90

BEG   MID   END

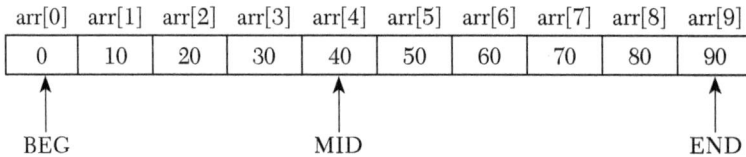

As ARR[5] = 50 > VAL = 30, therefore we will now search for the value in the lower half of the array. So now the values of END and MID are modified, and we move to the next pass.

Pass 2:

Now, $\text{END} = \text{MID} - 1 = 4$

$$\text{MID} = (0 + 4)/2 = 2$$

Now VAL = 20 and ARR[MID] = ARR[2] = 20.

arr[0]	arr[1]	arr[2]	arr[3]	arr[4]	arr[5]	arr[6]	arr[7]	arr[8]	arr[9]
0	10	20	30	40	50	60	70	80	90

BEG   MID   END

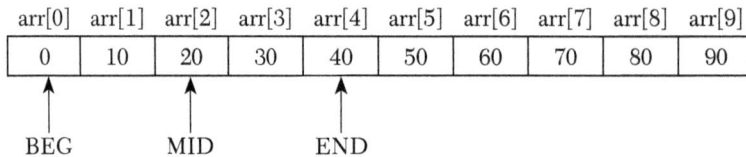

Figure 6.2. Working of a binary search.

Hence, the search is successful and VAL = 20 is found at POS = 2.

6.3.2 Complexity of a Binary Search Algorithm

In a binary search algorithm, we can see that with each comparison, the size of the search area is reduced by half. So, we can claim that the efficiency of the binary search in the worst case is $O(\log_{10}n)$, where n is the total number of elements in the array. Obviously, the best case will happen when the value to be searched for is equal to the value of the array in the middle.

6.3.3 Drawbacks of a Binary Search

- A binary search requires that the data elements in the array be sorted; otherwise, a binary search will not work.

■ A binary search cannot be used where there are many insertions and deletions of data elements in the array.

//Write a program to search for an element in an array using the binary search technique.

```java
import java.util.Scanner;
public class BinarySearchDemo {
    static Scanner scn = new Scanner(System.in);
    public static void main(String[] args) {
        int[] array = takeInput();
        System.out.print("\nEnter value to search:");
        int val = scn.nextInt();
        int pos = binary_search(array, val);
        if (pos < 0) {
            System.out.print("\n" + val + " not found");
        } else {
            System.out.print("\n" + val + " fount at position " +
                                          (pos + 1));
        }
    }
    public static int binary_search(int[] array, int value) {
        int low = 0, high = array.length - 1;
        while(low <=high) {
            int mid = low + high / 2;
            if (value< array[mid]) {
                high = mid - 1;

            } else if (value> array[mid]) {
                low = mid + 1;

            } else {
                return mid;
            }
        }
        return -1;
    }
    public static int[] takeInput() {
        System.out.print("\nEnter no of elements of array:");
        int size = scn.nextInt();
        int[] arr = new int[size];
        System.out.print("\nEnter the elements of array:");
        System.out.println();
        for (int i = 0; i < arr.length; i++) {
            System.out.print("Enter element "+(i+1)+": ");
```

```
        arr[i] = scn.nextInt();
    }
    return arr;
  }
}
```

The output of the program is shown as:

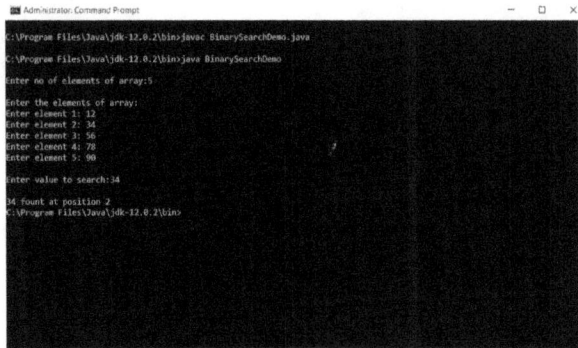

Frequently Asked Questions

2. What is a binary search? Explain.

Ans: *A binary search is one of the searching techniques which is used to find an element in an array. It works very efficiently with a sorted list. In a binary search, the element to be searched for is compared with the middle element of the array. If the value to be searched for is less than the middle element, we will search in the lower half of the array and vice versa.*

6.4 Interpolation Search

An interpolation search, also known as an extrapolation search, is a technique for searching for a particular value in an ordered array. This searching technique is more efficient than a binary search if the elements in the array are sorted. The technique of an interpolation search is similar to when we are searching for "Abbey" in the telephone directory; we don't start in the middle, because we know that it will be near the extreme left, so we start from the front and work from there. That is the main idea of an interpolation search; that is, instead of dividing the list into fixed halves, we cut it by an amount that seems most likely to succeed.

Practical Application

If we want to search for "Adda" in the directory, then we will always search in the extreme left of the directory.

6.4.1 Working of the Interpolation Search Algorithm

In each step of this searching technique, the remaining search area for the value to be searched for is calculated. The calculations are done on the values at the bounds of the search area and the value which is to be searched. Therefore, the value found at this position will now be compared with the value to be searched. If both values are equal, then the search is said to be successful. If both values are unequal, then depending upon the comparison done, the remaining search area is reduced to the part just before or after the initial position.

Consider an array ARR of n elements in which the elements are arranged in a sorted manner. Initially, low is set to 0 and high is set to n-1. Now we are searching for a value VAL in ARR between ARR[LOW] and ARR[HIGH]. Then, in this case, MID will be calculated by the following formula:

$$\textbf{MID} = \text{LOW} + (\text{HIGH} - \text{LOW}) \text{ X } ((\text{VAL} - \text{ARR[LOW]} / \text{ARR[HIGH]} - \text{ARR[LOW]}))$$

If the value VAL is found at MID, then the search is complete; otherwise, if the value is lower than ARR[MID], reset HIGH = MID – 1, and if the value is greater than ARR[MID], reset LOW = MID + 1. Repeat these steps until the value is found.

Hence, we can say that the interpolation search is very similar to the binary search technique. The main difference between the techniques is that in a binary search the value selected is always the middle value of the list, and it discards half the values based on the comparison between the value to be searched for and the value found at the estimated position. Let us understand the interpolation search with the help of an algorithm:

Algorithm for an Interpolation Search

INTERPOLATION_SEARCH(ARR, Lower_bound, Upper_bound, VAL)

```
Step 1: START
Step 2: Set LOW = lower_bound, HIGH = upper_bound, POS = -1
Step 3: Repeat Steps 4 & 5 while LOW<= HIGH
Step 4: Set MID = LOW + (HIGH - LOW) X ((VAL - ARR[LOW] /
                                    ARR[HIGH] - ARR[LOW]))
Step 5: IF (ARR[MID] = VAL)
          POS = MID
          PRINT POS
              Go to Step 7
          ELSE IF (ARR[MID] > VAL)
```

```
          Set HIGH = MID - 1
          ELSE
          Set LOW = MID + 1
          [End of If]
          [End of Loop]
Step 6:   IF (POS = -1)
          PRINT "VALUE NOT FOUND, SEARCH UNSUCCESSFUL"
          [End of IF]
Step 7:   EXIT
```

For example: Consider an array of seven numbers which is declared as:

```
int array[] = {5, 16, 23, 34, 45, 56, 65} ;
```

and the value to be searched for is 45.

Solution

Pass 1:

LOW = 0, HIGH = 7 − 1= 6, VAL =45

ARR[LOW] = ARR[0] = 5, ARR[HIGH] = ARR[6] = 65

Now, MID = LOW + (HIGH − LOW) × ((VAL − ARR[LOW]) /
$$(ARR[HIGH] − ARR[LOW]))$$

$$= 0 + (6 − 0) × ((45 − 5)/(65 − 5)$$

$$= 0 + 6 × (40/60) = 4$$

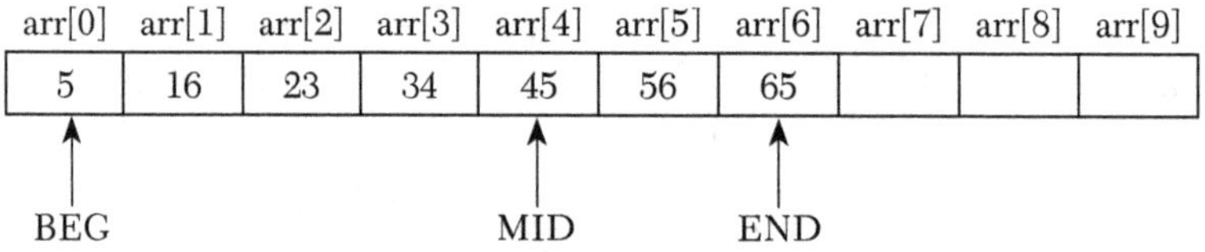

arr[0]	arr[1]	arr[2]	arr[3]	arr[4]	arr[5]	arr[6]	arr[7]	arr[8]	arr[9]
5	16	23	34	45	56	65			

BEG   MID   END

If(VAL == ARR[MID]) i.e. 45 == ARR[4] =45, 45 = 45

Figure 6.3. Working of the interpolation search.

Hence, the value is found.

6.4.2   Complexity of the Interpolation Search Algorithm

The interpolation search makes about $log_{10}(log_{10} n)$ *comparisons* when there are n elements in the list and the elements are uniformly distributed. Obviously, the worst case will happen when the number of elements is increased exponentially; in that case, the algorithm can take up to O(n) comparisons.

//Write a program to search for an element in an array using the interpolation search technique.

```java
import java.util.Scanner;
public class InterpolationSearchDemo {
    static Scanner scn = new Scanner(System.in);
    public static void main(String[] args) {
        int[] array = takeInput();
        System.out.print("\nEnter value to search:");
        int val = scn.nextInt();
        int pos = interpolation_search(array, val);
        if (pos < 0) {
            System.out.print("\n" + val + " not found");
        } else {
            System.out.print("\n" + val + " fount at position " +
                                            (pos + 1));
        }
    }
    public static int interpolation_search(int[] array, int value) {
        int low = 0, high = array.length - 1;
        while (low <= high) {
            int mid = low + ((value - array[low]) * (high - low))/
                                    (array[high] - array[low]);
            if (array[mid] == value) {
                return mid;
            } else if (array[mid] < value) {
                low = mid + 1;
            } else {
                high = mid - 1;
            }
        }
        return -1;
    }
    public static int[] takeInput() {
        System.out.print("\nEnter no of elements of array:");
        int size = scn.nextInt();
        int[] arr = new int[size];
        System.out.print("\nEnter the elements of array:");
        System.out.println();
        for (int i = 0; i < arr.length; i++) {
            System.out.print("Enter element " + (i + 1) + ": ");
            arr[i] = scn.nextInt();
        }
        return arr;
    }
}
```

The output of the program is shown as:

6.5 Introduction to Sorting

Sorting refers to the process of arranging the data elements of an array in a specified order, that is, either in ascending or descending order. For example, it will be practically impossible for us to find a name in the telephone directory if the names in it are not in alphabetical order. However, the same can be true for dictionaries, book indexes, bank accounts, and so on. Hence, the convenience of having sorted data is unquestionable. Retrieval of information becomes much easier when the data is stored in some specified order. Therefore, sorting is a very important application in computer science.

Let us take an array which is declared and initialized as:

```
int array[] = {10, 25, 17, 8, 30, 3} ;
```

Then, the array after applying the sorting technique is:

```
array[] = {3, 8, 10, 17, 25, 30} ;
```

A sorting algorithm can be defined as an algorithm which puts the data elements of an array/ list in a certain order, that is, either numerical order or any predefined order. There are many sorting algorithms which are available and are widely used according to the different environments required by the different sorting methods.

The two basic categories of sorting methods are:

1. **Internal Sorting:** It refers to the sorting of the data elements stored in the computer's main memory.

2. **External Sorting:** It refers to the sorting of the data elements stored in the files. It is applied when the amount of data is large and cannot be stored in the main memory.

6.5.1 Types of Sorting Methods

The various sorting methods are:

1. Selection Sort

2. Insertion Sort

3. Merge Sort

4. Bubble Sort

5. Quick Sort

Let us discuss all of them in detail.

1. Selection Sort

Selection sort is a sorting technique that works by finding the smallest value in the array and placing it in the first position. After that, it then finds the second smallest value and places it in the second position. This process is repeated until the whole array is sorted. Thus,the selection sort works by finding the smallest unsorted element remaining in the entire array and then swapping it with the element in the next position to be filled. It is a very simple technique, and it is also easier to implement than other sorting techniques. Selection sort is used for sorting files with large records.

Selection Sort Technique

Let us take an array ARR with N elements in it. Now, the selection sort technique works as follows:

First of all, we will find the smallest value in the entire array, and we will place that value in the first position of the array. Then, we will find the second smallest value in the array, and we will place it in the second position of the array. Now, we will repeat this process until the whole array is sorted.

Pass 1: Find the position POS of the smallest value in the array of N elements and interchange ARR[POS] with ARR[0]. Hence, ARR[0] is sorted.

Pass 2: Find the position POS of the smallest value in the array of N-1 elements and interchange ARR[POS] with A[1]. Hence, A[1] is sorted.

.

.

.

Pass N-1: Find the position POS of the smaller of the elements of ARR[N-2] and ARR[N-1] and interchange ARR[POS] with ARR[N-2]. Hence, ARR[0], ARR[1], . . . ARR[N-1] is sorted.

Let us discuss it with the help of a detailed algorithm.

Algorithm for a Selection Sort

Consider an array ARR having N elements from ARR[0] to ARR[N-1]. I and J are the looping variables, and POS is the swapping variable.

SELECTION SORT(ARR, N)

```
Step 1: START
Step 2: Repeat Steps 3 & 4 for I = 1 to N - 1
Step 3: Call MIN(ARR, I, N, POS)
Step 4: Swap ARR[I] with ARR[POS]
        [End of Loop]
Step 5: EXIT
        MIN(ARR, I, N, POS)
Step 1:Set SMALLEST = ARR[I]
Step 2:Set POS = I
Step 3: Repeat Step 4 for J = I + 1 to N - 1
Step 4: IF (ARR[J] < SMALLEST)
            Set SMALLEST = ARR[J]
        Set POS = J
        [End of IF]
        [End of Loop]
Step 5: Return POS
```

For Example: Sort the given array using selection sort.

arr[0]	arr[1]	arr[2]	arr[3]	arr[4]
4	14	29	11	35

Solution:

Pass	POS	Array[0]	Array[1]	Array[2]	Array[3]	Array[4]
1	4	4	14	29	11	35
2	3	4	11	29	14	35
3	3	4	11	14	29	35
4	3	4	11	14	29	35
5	4	4	11	14	29	35

Figure 6.4. Working of selection sort.

Hence, after sorting the new array is:

arr[0]	arr[1]	arr[2]	arr[3]	arr[4]
4	11	14	29	35

Complexity of the Selection Sort Algorithm

Selection sort is the simple technique of sorting. In this method, if there are n elements in the array, then (n-1) comparisons or iterations are made. Thus, *the selection sort technique has a complexity of $O(n^2)$.*

//Write a program to sort an array using the selection sort method.

```
import java.util.Scanner;
public class SelectionSortDemo {
    static Scanner scn = new Scanner(System.in);
    public static void main(String[] args) {
        int[] array = takeInput();
        selection_sort(array);
        display(array);
    }
    public static int[] takeInput() {
        System.out.print("\nEnter no of elements of array:");
        int size = scn.nextInt();
        int[] arr = new int[size];
        System.out.print("\nEnter the elements of array:");
        System.out.println();
        for (int i = 0; i < arr.length; i++) {
            System.out.print("Enter element "+(i+1)+": ");
            arr[i] = scn.nextInt();
        }
        return arr;
    }
    public static void selection_sort(int[] array) {
        int n = array.length - 2;
        for (int c = 0; c <= n; c++) {
            int min=c;
            for (int j = c+1; j <= n+1; j++) {
                if (array[j]<array[min]) {
                    min=j;
                }
            }
            int temp = array;
            array = array[min];
            array[min] = temp;
        }
    }
}
```

```java
public static void display(int[] arr) {
    System.out.print("\nAfter sorting array is: ");
    for (int val : arr) {
        System.out.print(val + " ");
    }
}
}
```

The output of the program is shown as:

Frequently Asked Questions

3. Define the selection sort technique.

Ans: *Selection sort is a sorting technique which works by finding the smallest element from the array and placing it in the first position. It then finds the second smallest element and places it in the second position. Hence, this procedure is repeated until the whole array is sorted.*

---

2. Insertion Sort

Insertion sort is another very simple sorting algorithm which works just like its name suggests; that is, it inserts each element into its proper position in the concluding list. To limit the wastage of memory or, we can say, to save memory, most implementations of an insertion sort work by moving the current element past the already sorted elements and repeatedly swapping or interchanging it with the preceding element until it is placed in its correct position.

Practical Application

We usually use this technique while ordering a deck of cards while playing a game called bridge.

Insertion Sort Technique

Pass 1: Initially there is only one element in the list which is already sorted. Hence, we proceed to the next steps.

Pass 2: During the first iteration, the first and the second element of the list are compared. The smaller value occupies the first position of the list.

Pass 3: During the second iteration, the first three elements of the list are compared. The smaller value will occupy the first position in the list. The second position will be occupied by the second smallest element, and so on.

This procedure is repeated for all the elements of the array up to (n-1) iterations.

Algorithm for an Insertion Sort

INSERTION SORT(ARR, N)

```
Step 1: START
Step 2: Repeat Steps 3 to 6 for I = 1 to N - 1
Step 3: Set POS = ARR[I]
Step 4: Set J = I - 1
Step 5: Repeat while POS <= ARR[J]
            Set ARR[J + 1] = ARR[J]
            Set J = J - 1
        [End of Inner while loop]
Step 6: Set ARR[J + 1] = POS
        [End of Loop]
Step 7: EXIT
```

In the previous algorithm, in Step 2, a for loop is executed which will be repeated for every element in the array. In Step 3, we are storing the value of the Ith element in POS. In Step 5, again a loop is executed in which the new elements after sorting are placed. At last, the element is stored at the (J+1)th position.

For example: Consider the following array. Sort the given values in the array using the insertion sort technique.

arr[0]	arr[1]	arr[2]	arr[3]	arr[4]	arr[5]
39	54	10	28	95	7

Solution:

Sorted	Unsorted

Pass 1: Initially, ARR[0] is sorted. Move to the next pass.

39	54	10	28	95	7

Pass 2: Now 39 and 54 are compared. 39 < 54, so ARR[0] = 39 and ARR[1] = 54.

39	54	10	28	95	7

Pass 3: 39, 54, and 10 are compared. 10 < 39 and 54, so ARR[0] = 10, now 39 < 54 hence ARR[1] = 39 and ARR[2] = 54.

10	39	54	28	95	7

Pass 4: As 28 < 39 and 54, so ARR[1] = 28.

10	28	39	54	95	7

Pass 5: In this case, 95 is greater than all the values, so there is no need for swapping.

10	28	39	54	95	7

Pass 6: 7 is the smallest value, so ARR[0] = 7.

Therefore, after sorting the new array is:

7	10	28	39	54	95

Figure 6.5. Working of an insertion sort.

Complexity of an Insertion Sort

In an insertion sort, the best case will happen when the array is already sorted, and in that case the running time of the algorithm is O(n)(i.e., linear running time).Obviously, the worst case will happen when the array is sorted in the reverse order. Thus, in that case the running time of the algorithm is O(n²) (i.e., quadratic running time).

//Write a program to sort an array using the insertion sort method.

```java
import java.util.Scanner;
public class InsertionSortDemo {
    static Scanner scn = new Scanner(System.in);
    public static void main(String[] args) {
        int[] array = takeInput();
        insertion_sort(array);
        display(array);
    }
    public static int[] takeInput() {
        System.out.print("\nEnter no of elements of array:");
```

```
        int size = scn.nextInt();
        int[] arr = new int[size];
        System.out.print("\nEnter the elements of array:");
        System.out.println();
        for (int i = 0; i < arr.length; i++) {
            System.out.print("Enter element "+(i+1)+": ");
            arr[i] = scn.nextInt();
        }
        return arr;
    }
    public static void insertion_sort(int[] array) {
        int n = array.length;
        for (int c = 0; c < n; c++) {
            int temp = array;
            int j = c - 1;
            while (j >= 0 && array[j] > temp) {
                array[j + 1] = array[j];
                j--;
            }
            array[j + 1] = temp;
        }
    }
    public static void display(int[] arr) {
        System.out.print("\nAfter sorting array is: ");
        for (int val : arr) {
            System.out.print(val + " ");
        }
    }
}
```

The output of the program is shown as:

3. Merge Sort

Merge sort is a sorting method which follows the divide and conquer approach. The divide and conquer approach is a very good approach in which divide means partitioning the array having n elements into two sub-arrays of n/2 elements each. However, if there are no elements present in the list/array or if an array contains only one element, then it is already sorted. However, if an array has more elements, then it is divided into two sub-arrays containing equal elements in them. Conquer is the process of sorting the two sub-arrays recursively using merge sort. Finally, the two sub-arrays are merged into one single sorted array.

Merge Sort Techniques

1. If the array has zero or one element in it, then there is no need to sort that array as it is already sorted.

2. Otherwise, if there are more elements in the array, then divide the array into two sub-arrays containing equal elements.

3. Each sub-array is now sorted recursively using merge sort.

4. Finally, the two sub-arrays are merged into a single sorted array.

Algorithm of Merge Sort

MERGE SORT(ARR, BEG, END)

```
Step 1: START
Step 2: IF (BEG < END)
Step 3: Set MID = (BEG + END)/2
        Call MERGE SORT (ARR, BEG, MID)
        Call MERGE SORT (ARR, MID + 1, END)
        Call MERGE (ARR, BEG, MID, END)
        [End ofIf]
Step 4: EXIT
MERGE(ARR, BEG, MID, END)
Step 1: START
Step 2: Set I = BEG, J = MID + 1, K = 0
Step 3: Repeat while (I <= MID)  && (J <= END)
        IF (ARR[I] > ARR[J])
              Set TEMP[K] = ARR[J]
              Set J = J + 1
              Set K = K + 1
        ELSE IF (ARR[J] > ARR[I])
              Set TEMP[K] = ARR[I]
              Set I = I + 1
              Set K = K + 1
        ELSE
              Set TEMP[K] = ARR[J]
              Set J = J + 1
```

```
            Set K = K + 1
            Set TEMP[K] = ARR[I]
            Set I = I + 1
            Set K = K + 1
        [End of If]
        [End of Loop]
Step 4: (Copying the remaining elements of left sub array if any)
        Repeat while (I <= MID)
            Set TEMP[K] = ARR[I]
            Set I = I + 1
            Set K = K + 1
        [End of Loop]
Step 5: (Copying the remaining elements of right sub array if any)
        Repeat while (J <= END)
            Set TEMP[K] = ARR[J]
            Set I = I + 1
            Set K = K + 1
        [End of Loop]
Step 6: Set IND = 0
Step 7: Repeat while (IND < K)
        Set ARR[IND] = ARR[IND]
            Set IND = IND + 1
        [End of Loop]
Step 8: EXIT
```

For example: Sort the following array using merge sort.

```
        int array[] = { 40, 10, 86, 44, 93, 26, 69, 17 }
```

Solution:

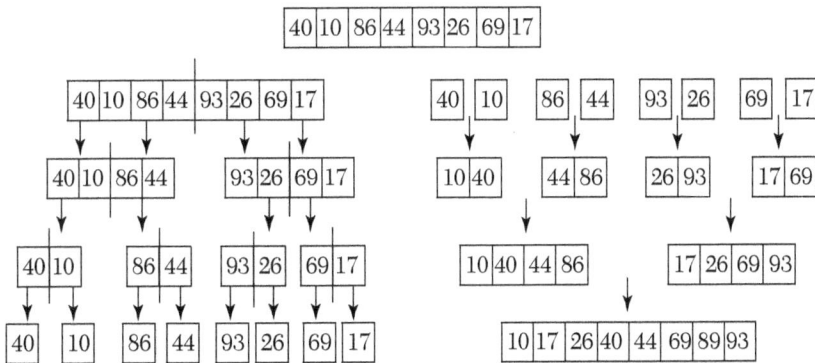

Divide and Conquer Process   Merging the sub arrays into one sorted array

Figure 6.6. Working of a merge sort.

From the previous example, we can see how the merge sort algorithm works. First, the merge sort algorithm recursively divides the array into smaller sub-arrays. After dividing the array into smaller parts, we call the function Merge() to merge all the sub-arrays to form a single sorted array.

Complexity of Merge Sort

The running time of the merge sort algorithm is $O(n \log_{10} n)$. This runtime remains the same in the average as well as in the worst case of the merge sort algorithm. Although it has an optimal time complexity, sometimes this runtime can be $O(n)$.

//Write a program to sort an array using the merge sort method.

```java
import java.util.Scanner;
public class MergeSortDemo {
    static Scanner scn = new Scanner(System.in);
    public static void main(String[] args) {
        int[] array = takeInput();
        int[] ans = merge_sort(array, 0, array.length - 1);
        display(ans);
    }
    public static int[] takeInput() {
        System.out.print("\nEnter no of elements of array:");
        int size = scn.nextInt();
        int[] arr = new int[size];
        System.out.print("\nEnter the elements of array:");
        System.out.println();
        for (int i = 0; i < arr.length; i++) {
            System.out.print("Enter element " + (i + 1) + ": ");
            arr[i] = scn.nextInt();
        }
        return arr;
    }
    public static int[] merge_sort(int[] arr, int lo, int hi) {
        if (lo == hi) {
            int[] br = new int[1];
            br[0] = arr[lo];
            return br;
        }
        int mid = (lo + hi) / 2;
        int[] fh = merge_sort(arr, lo, mid);
        int[] sh = merge_sort(arr, mid + 1, hi);
        int[] merged = merge_sorted_arrays(fh, sh);
        return merged;
    }
```

```java
public static int[] merge_sorted_arrays(int[] one, int[] two) {
    int[] ans = new int[one.length + two.length];
    int i = 0;
    int j = 0;
    int k = 0;
    while (i < one.length && j < two.length) {
        if (one[i] < two[j]) {
            ans[k] = one[i];
            i++;
            k++;
        } else {
            ans[k] = two[j];
            j++;
            k++;
        }
    }
    if (i == one.length) {
        while (j < two.length) {
            ans[k] = two[j];
            j++;
            k++;
        }
    }
    if (j == two.length) {
        while (i < one.length) {
            ans[k] = one[i];
            i++;
            k++;
        }
    }
    return ans;
}
public static void display(int[] arr) {
    System.out.print("\nAfter sorting array is: ");
    for (int val : arr) {
        System.out.print(val + " ");
    }
}
}
```

The output of the program is shown as:

4. Bubble Sort

Bubble sort, also known as exchange sort, is a very simple sorting method. It works by repeatedly moving the largest element to the highest position of the array. In bubble sort, we are comparing two elements at a time, and swapping is done if they are wrongly placed. If the element at a lower index or position is greater than the element at a higher index, then in that case both the elements are interchanged so that the smaller element is placed before the bigger one. This process is repeated until the list becomes sorted. Bubble sort gets its name from the way that the smaller elements "bubble" to the top of the array. This sorting technique only uses comparisons to operate on the elements. Hence, we can also call it a comparison sort.

Bubble Sort Technique

The basic idea applied for a bubble sort is to let us assume if an array ARR contains n elements, then the number of iterations required to sort the array will be (n − 1).

Pass 1: During the first iteration, the largest value in the array is placed at the last position.

Pass 2: During the second iteration, the second largest value of the array is placed in the second to last position.

Pass 3: During the third iteration, the third largest value of the array is placed in the third to last position, and so on.

This procedure is repeated until all the elements in the array are scanned and are placed in their correct position, which means that the array is sorted.

Algorithm of a Bubble Sort

```
BUBBLE SORT(ARR, N)
Step 1: START
Step 2: Repeat Step 3 for I = 0 to N - 1
Step 3: Repeat for J = 0 to N - 1
Step 4: IF (ARR[J] > ARR[J+1])
            INTERCHANGE ARR[J] & ARR[J + 1]
        [End of Inner Loop]
        [End of Outer Loop]
Step 5: EXIT
```

For example: Consider the following array. Sort the given values in the array using the bubble sort technique.

arr[0]	arr[1]	arr[2]	arr[3]	arr[4]
40	50	20	90	30

Solution: In the given array, the number of elements in the array is 5, so the number of iterations will be $(n - 1) = 4$.

Pass 1:

40	50	20	90	30

(a) 40 and 50 are compared. Since 40 < 50, no swapping is done.

40	50	20	90	30

(b) 50 and 20 are compared. Since 50> 20, swapping will be done.

40	20	50	90	30

(c) 50 and 90 are compared. Since 50 < 90, no swapping is done.

40	20	50	90	30

(d) 90 and 30 are compared. Since 90 > 30, swapping is done.

40	20	50	30	90

At the end of the first pass, the largest element in the array is placed at the highest position in the array, but all the other elements are still unsorted. Let us now proceed to Pass 2.

Pass 2:

40	20	50	30	90

(a) 40 and 20 are compared. Since 40 > 20, swapping is done.

20	40	50	30	90

(b) 40 and 50 are compared. Since 40 < 50, no swapping will be done.

20	40	50	30	90

(c) 50 and 30 are compared. Since 50 > 30, swapping is done.

20	40	30	50	90

At the end of the second pass, the second largest element in the array is placed at the second last position in the array, but all the other elements are still unsorted. Let us now proceed to Pass 3.

Pass 3:

20	40	30	50	90

(a) 20 and 40 are compared. Since 20 < 40, no swapping is done.

20	40	30	50	90

(b) 40 and 30 are compared. Since 40 > 30, swapping will be done.

20	30	40	50	90

At the end of the third pass, the third largest element in the array is placed at the third largest position in the array, but all the other elements are still unsorted. Let us now proceed to Pass 4.

Pass 4:

20	40	30	50	90

(a) 20 and 40 are compared. Since 20 < 40, no swapping is done.

At the end of the fourth pass, we can see that all the elements in the list are sorted. Hence, after sorting, the new array will be:

20	40	30	50	90

Figure 6.7. Working of the bubble sort.

Complexity of the Bubble Sort

In the best case, the running time of the bubble sort is O(n), that is, when the array is already sorted. Otherwise, its level of complexity in average and worst cases is $O(n^2)$.

//Write a program to sort an array using the bubble sort method.

```java
import java.util.Scanner;
public class BubbleSortDemo {
    static Scanner scn = new Scanner(System.in);
    public static void main(String[] args) {
        int[] array = takeInput();
        bubble_sort(array);
        display(array);
    }
    public static int[] takeInput() {
        System.out.print("\nEnter no of elements of array:");
        int size = scn.nextInt();
        int[] arr = new int[size];
        System.out.print("\nEnter the elements of array:");
        System.out.println();
        for (int i = 0; i < arr.length; i++) {
            System.out.print("Enter element "+(i+1)+": ");
            arr[i] = scn.nextInt();
        }
        return arr;
    }
    public static void bubble_sort(int[] array) {
        int n = array.length - 2;
        for (int c = 0; c <= n; c++) {
            for (int j = 0; j <= n - c; j++) {
                if (array[j] > array[j + 1]) {
                    int temp = array[j];
                    array[j] = array[j + 1];
                    array[j + 1] = temp;
                }
            }
        }
    }
    public static void display(int[] arr) {
        System.out.print("\nAfter sorting array is: ");
        for (int val : arr) {
            System.out.print(val + " ");
        }
    }
}
```

The output of the program is shown as:

5. Quick Sort

Quick sort, also known as *partition exchange sort* and developed by C. A. R. Hoare, is a widely used sorting algorithm which also uses the divide and conquer approach, as we have discussed in merge sort. Here we will also divide a single unsorted array into its two smaller sub-arrays. The divide and conquer method means dividing the bigger problem into two smaller problems, and then those two smaller problems into smaller problems, and so on. Like merge sort, if there are no elements in the array or if an array contains only one element, then it is already sorted. A quick sort algorithm is faster than all the other sorting algorithms which have time complexity $O(n \log_{10} n)$.

Working of Quick Sort

1. An element called a pivot is selected from the array elements.

2. After choosing the pivot element, all the elements of the array are rearranged such that all the elements less than the pivot element will be on left side, and all the elements greater than the pivot element will be placed on the right side of the pivot element. After rearranging all the elements, the pivot is now placed in its final position. Thus, this process is known as partitioning.

3. Now, the two sub-arrays obtained will be recursively sorted.

Quick Sort Technique

1. Initially set the index of the first element to LEFT and POS. Similarly, set the index of the last element to RIGHT. Now, LEFT = 0, POS = 0, RIGHT = N − 1 (assuming n elements in the array).

2. We will start with the last element, which is pointed to by RIGHT, and we will traverse each element in the array from right to left, comparing each element with the first element pointed to by POS. ARR[POS] should always be less than ARR[RIGHT].

- If ARR[POS] is less than ARR[RIGHT], then continue comparing until RIGHT = POS. If RIGHT = POS then it means that the pivot is placed in its correct position.

- If ARR[RIGHT] < ARR[POS], then swap the two values and go to the next step.

- Set POS = RIGHT.

3. We will start from the first element, which is pointed to by LEFT, and we will traverse every element in the array from left to right, comparing each element with the element pointed to by POS. ARR[POS] should always be greater than ARR[LEFT].

- If ARR[POS] is greater than ARR[RIGHT], then continue comparing until LEFT = POS. If LEFT = POS then it means that the pivot is placed in its correct position.

- If ARR[LEFT] > ARR[POS], then swap the two values and go to the previous step.

- Set POS = LEFT.

Algorithm of Quick Sort

QUICK SORT(ARR, BEG, END)

```
Step 1: START
Step 2: IF (BEG < END)
            Call PARTITION (ARR, BEG, END, POS)
            Call QUICK SORT (ARR, BEG, POS - 1)
        Call QUICK (ARR, POS + 1, END)
            [End of If]
Step 3: EXIT
```

PARTITION(ARR, BEG, END, POS)

```
Step 1: START
Step 2: Set LEFT = BEG, RIGHT = END, POS = BEG, TEMP = 0
Step 3: Repeat Steps 4 to 7 while TEMP = 0
Step 4: Repeat while ARR[RIGHT] >= ARR[POS]&& POS != RIGHT
            Set RIGHT = RIGHT - 1
            [End of Loop]
```

```
Step 5: IF (POS = RIGHT)
            Set TEMP = 1
            ELSE IF (ARR[POS] > ARR[RIGHT])
                INTERCHANGE ARR[POS] with ARR[RIGHT]
                Set POS = RIGHT
            [End of If]
Step 6: IF TEMP = 0
                Repeat while ARR[POS] >= ARR[LEFT] && POS != LEFT
                Set LEFT = LEFT + 1
            [End of Loop]
Step 7: IF (POS = LEFT)
            Set TEMP = 1
ELSE IF (ARR[LEFT] > ARR[POS])
                INTERCHANGE ARR[POS] with ARR[LEFT]
                Set POS = LEFT
            [End of If]
            [End of If]
            [End of Loop]
Step 8: EXIT
```

For example: Sort the values given in the following array using the quick sort algorithm.

arr[0]	arr[1]	arr[2]	arr[3]	arr[4]	arr[5]
25	7	39	17	30	52

Solution:

Step 1: The first element is chosen as the pivot. Now, set POS = 0, LEFT = 0, RIGHT = 5.

25	7	39	17	30	52

POS, LEFT   RIGHT

Step 2: Traverse the list from right to left. Since ARR[POS] < ARR[RIGHT],that is, 25 < 52,   RIGHT = RIGHT – 1 = 4.

25	7	39	17	30	52

POS, LEFT   RIGHT

Step 3: Since ARR[POS] < ARR[RIGHT],that is, 25 < 30, RIGHT = RIGHT – 1 = 3.

25	7	39	17	30	52

POS, LEFT   RIGHT

Step 4: Since ARR[POS] > ARR[RIGHT], that is, 25 > 17, we will swap the two values and set POS = RIGHT.

17	7	39	25	30	52

LEFTRIGHT, POS

Step 5: Traverse the list from left to right. Since ARR[POS] > ARR[LEFT],that is, 25 > 17, LEFT = LEFT + 1.

17	7	39	25	30	52

LEFT   RIGHT, POS

Step 6: Since ARR[POS] > ARR[LEFT],that is, 25 > 7, LEFT = LEFT + 1.

17	7	39	25	30	52

LEFT   RIGHT, POS

Step 7: Since ARR[POS] < ARR[LEFT],that is, 25 < 39, we will swap the values and set POS = LEFT.

17	7	25	39	30	52

LEFT, POS RIGHT

Step 8: Traverse the list from right to left. Since ARR[POS] < ARR[LEFT], RIGHT = RIGHT - 1.

17	7	25	39	30	52

LEFT, POS, RIGHT

Now, RIGHT = POS, so now the process is over and the pivot element of the array, that is, 25, is placed in its correct position. Therefore, all the elements which are smaller than 25 are placed before it, and all the elements greater than 25 are placed after it. Hence, 17 and 7 are the elements in the left sub-array, and 39, 30, and 52 are the elements in the right sub-array, which both are sorted.

Figure 6.8. Working of quick sort.

Complexity of Quick Sort

The running time efficiency of quick sort is $O(n \log_{10} n)$ in the average and the best case. However, the worst case will happen if the array is already sorted and the leftmost element is selected as the pivot element. In the worst case, its efficiency is $O(n^2)$.

//Write a program to sort an array using the quick sort method.

```java
import java.util.Scanner;
public class QuickSortDemo {
    static Scanner scn = new Scanner(System.in);
    public static void main(String[] args) {
        int[] array = takeInput();
        quick_sort(array, 0, array.length - 1);
        display(array);
    }
    public static int[] takeInput() {
        System.out.print("\nEnter no of elements of array:");
        int size = scn.nextInt();
        int[] arr = new int[size];
        System.out.print("\nEnter the elements of array:");
        System.out.println();
        for (int i = 0; i < arr.length; i++) {
            System.out.print("Enter element " + (i + 1) + ": ");
            arr[i] = scn.nextInt();
        }
        return arr;
    }
    public static void quick_sort(int[] arr, int lo, int hi) {
        if (lo >= hi) {
            return;
        }
        int mid = (lo + hi) / 2;
        int pivot = arr[mid];
        int left = lo;
        int right = hi;
        while (left <= right) {
            while (arr[left] < pivot) {
                left++;
            }
            while (arr[right] > pivot) {
                right--;
            }
            if (left <= right) {
                int temp = arr[left];
                arr[left] = arr[right];
                arr[right] = temp;
                left++;
                right--;
            }
        }
```

```
        }
        quick_sort(arr, lo, right);
        quick_sort(arr, left, hi);
    }
    public static void display(int[] arr) {
        System.out.print("\nAfter sorting array is: ");
        for (int val : arr) {
            System.out.print(val + " ");
        }
    }
}
```

The output of the program is shown as:

6.6 External Sorting

External sorting is a sorting technique which is used when the amount of data is massive. When a large amount of data has to be sorted, it is not possible to bring it into main memory (RAM). Therefore, in that situation a secondary memory needs to be used. Also, at the same time, some portion of data is brought into the main memory from the secondary memory for sorting based on the availability of storage space in the main memory. After the data is sorted, it is sent back to the secondary memory. Now, the next portion of the data is brought into the main memory, and after sorting it is sent back to the secondary memory. This procedure is repeated until all the data is sorted. Here, each portion is called a segment. The time required for sorting is greater because time will be spent transferring the data from secondary memory to main memory. The merge sort algorithm is widely and commonly used in external sorting, which has already been discussed.

External sorting is used in database applications for performing different kinds of operations like join, union, projection, and many more. It is also used to update a master file from a transaction file; for example, if we are updating the company file based on new employees, existing employees, locations, and so on. Duplicate records or data can also be removed from external sorting.

6.7 Summary

- The process of finding a particular value in a list or an array is called searching. If that particular value is present in the array, then the search is said to be successful, and the location of that particular value is retrieved by the searching process.

- Linear search, binary search, and interpolation search are the commonly used searching techniques.

- Linear search works by comparing the values to be searched for with every element of the array in a linear sequence until a match is found.

- Binary search works efficiently when the list is sorted. In a binary search, we first compare the value VAL with the data element in the middle position of the array.

- Interpolation search, also known as extrapolation search, is a technique for searching for a particular value in an ordered array. In each step of this searching technique, the remaining search area for the value to be searched for is calculated. The calculations are done on the values at the bounds of the search area and the value which is to be searched.

- Sorting refers to the technique of arranging the data elements of an array in a specified order, that is, either in ascending or descending order.

- Selection sort is a sorting technique that works by finding the smallest value in the array and placing it in the first position. After that, it then finds the second smallest value and places it in the second position. This process is repeated until the whole array is sorted.

- Insertion sort works by moving the current data element past the already sorted data elements and repeatedly interchanging it with the preceding element until it is in the correct place.

- Merge sort is a sorting method which follows the divide and conquer approach. Divide means partitioning the array having n elements into two sub-arrays of n/2 elements each. Conquer is the process of sorting the two sub-arrays recursively using merge sort. Finally, the two sub-arrays are merged into one single sorted array.

- Bubble sort, also known as exchange sort, is a very simple sorting method. It works by repeatedly moving the largest element to the highest position of the array.

- Quick sort is an algorithm which selects a pivot element and rearranges the values in such a way that all the elements less than the pivot element appear before it and the elements greater than the pivot appear after it.

- External sorting is a sorting technique which is used when the amount of data is massive.

6.8 Exercises

6.8.1 Theory Questions

1. Define sorting. Write the importance of sorting.

2. What are the different types of sorting techniques? Discuss each of them in detail.

3. Discuss the limitations and advantages of insertion sort.

4. Explain the working of bubble sort with a suitable example. Why is bubble sort called "bubble"?

5. Define searching. Which searching technique would you prefer while searching for an element in an array?

6. How is linear search used to find an element? Explain the working of insertion sort with a suitable example.

7. Explain different types of searching techniques. Give a suitable example to illustrate binary search.

8. Why is quick sort known as "quick"?

9. Explain the concept of external sorting.

10. Differentiate between binary search and interpolation search. Give a suitable example.

6.8.2 Programming Questions

1. Write a Java program to implement the bubble sort technique.

2. Write an algorithm to implement the interpolation search technique.

3. Write an algorithm to perform a merge sort. Show various stages in merge sorting over the data: 11, 2, 9, 13, 57, 25, 17, 1, 90, and 3.

4. Write a Java program to implement an insertion sort.

5. Write a program to search for an element using the binary search technique.

6. Write a Java program to perform a comparison sort.

7. Write an algorithm to perform a partition exchange sort technique. Show various stages over the data: 24, 52, 98, 12, 45, 6, 59, and 90.

8. Write an algorithm/program to implement a linear search technique.

6.8.3 Multiple Choice Questions

1. A binary search algorithm can be applied to a _____.
   (a) Sorted array
   (b) Sorted linked list
   (c) Unsorted linked list
   (d) Binary trees

2. The time complexity of a bubble sort algorithm is:
   (d) $O(\log n)$
   (b) $O(n)$
   (c) $O(n.\log n)$
   (d) $O(n^2)$

3. Which sorting algorithm is known as a partition exchange sort?
   (a) Selection Sort
   (b) Merge Sort
   (c) Quick Sort
   (d) Bubble Sort

4. Which case would exist when the element to be searched for using linear search is equal to the first element of the array?
   (a) Best Case
   (b) Worst Case
   (c) Average Case
   (d) None of these

5. Quick sort is faster than _____.

   (a) Bubble Sort

   (b) Selection Sort

   (c) Insertion Sort

   (d) All of the above

6. When the amount of data is massive, which type of sorting is preferred?

   (a) Internal Sorting

   (b) External Sorting

   (c) Both of these

   (d) None of these

7. Which of the searching techniques will be best when the value to be searched for is present in the middle?

   (a) Linear Search

   (b) Interpolation Search

   (c) Binary Search

   (d) All of these

8. The complexity of a binary search algorithm is _____.

   (a) $O(n^2)$

   (b) $O(\log n)$

   (c) $O(n)$

   (d) $O(n \log n)$

9. Selection sort has a linear running time complexity.

   (a) True

   (b) False

   (c) Not possible to comment

CHAPTER 7

STACKS

7.1 Introduction

A stack is an important data structure which is widely used in many computer applications. A stack can be visualized with many familiar examples from our day-to-day lives. A very simple illustration of a stack is a pile of books where one book is placed on top of another as in Figure 7.1.When we want to remove a book, we remove the topmost book first. Hence, we can add or remove an element (i.e., book) only at or from one position, which is the topmost position. There are many other daily life examples in which we can see how a stack is implemented. We observe that whenever we talk about a stack, we see that the element at the last position will be served first. Thus, a stack can be described as a LIFO (last in, first out) data structure; that is, the element which is inserted last will be the first one to be taken out. Now, let us discuss stacks in detail.

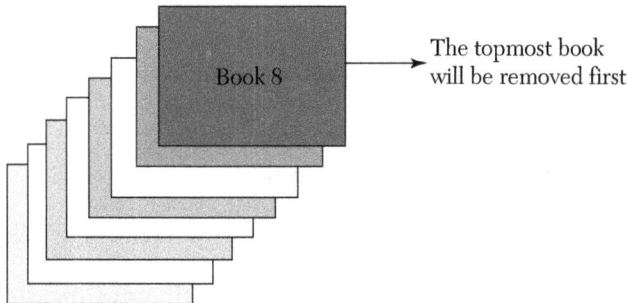

The topmost book will be removed first

Book 8

Figure 7.1. Stack of books.

7.2   Definition of a Stack

A stack is a linear collection of data elements in which the element inserted last will be the element taken out first (i.e., a stack is a LIFO data structure). The stack is an abstract data structure, somewhat similar to queues. Unlike queues, a stack is open only on one end. *The stack is a linear data structure in which the insertion and deletion of elements is done only from the end called TOP.* One end is always closed, and the other end is used to insert and remove data.

Stacks can be implemented by using arrays or linked lists. We will discuss the implementation of stacks using arrays and linked lists in this section.

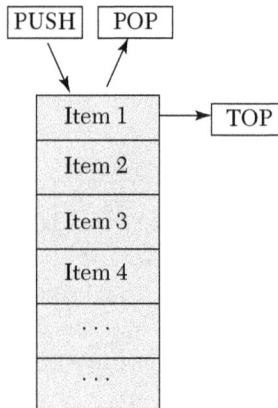

Figure 7.2. Representation of a stack.

Practical Application

1. A real-life example of a stack is a pile of dishes where one dish is placed on top of another. Now, when we want to remove a dish, we remove the topmost dish first.

2. Another real-life example of a stack is a pile of discs where one disc is placed on top of another. Now, when we want to remove a disc, we remove the topmost disc first.

7.3   Overflow and Underflow in Stacks

Let us discuss both overflow and underflow in stacks in detail:

1. **Overflow in stacks:** The overflow condition occurs when we try to insert elements in a stack, but the stack is already full. If an attempt is

made to insert a value in a stack that is already full, an overflow message is printed. It can be checked by the following formula:

If $TOP = MAX - 1$, where MAX is the size of the stack.

2. **Underflow in stacks:** The underflow condition occurs when we try to remove elements from a stack, but the stack is already empty. If an attempt is made to delete a value from a stack that is already empty, an underflow message is printed. It can be checked by the following formula:

If $TOP = NULL$, where MAX is the size of the stack.

Frequently Asked Questions

1. Define a stack and list the operations performed on stacks.

Ans: *A stack is a linear data structure in which the insertion and deletion of an element is done only from the end called TOP. It is LIFO in nature (i.e., Last In First Out). Different operations that can be performed on stacks are:*

(a) Push operation
(b) Pop Operation
(c) Peek Operation

7.4   Operations on Stacks

The three basic operations that can be performed on stacks are:

1. PUSH

The push operation is the process of adding new elements in the stack. However, before inserting any new element in the stack, we must always check for the overflow condition, which occurs when we try to insert an element in a stack which is already full. An overflow condition can be checked as follows, if TOP = MAX – 1,where MAX is the size of the stack. Hence, if the overflow condition is true, then an overflow message is displayed on the screen; otherwise, the element is inserted into the stack.

For example: Let us take a stack which has five elements in it. Suppose we want to insert another element, 10, in it; then TOP will be incremented by 1. Thus, the new element is inserted at the position pointed to by TOP. Now, let us see how a push operation occurs in the stack in the following figure:

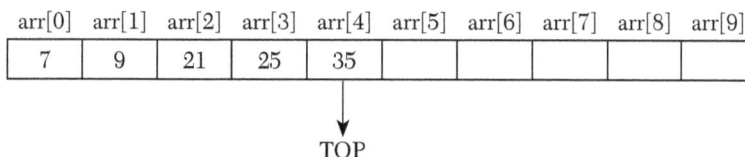

arr[0]	arr[1]	arr[2]	arr[3]	arr[4]	arr[5]	arr[6]	arr[7]	arr[8]	arr[9]
7	9	21	25	35					

TOP

After inserting 10 in it, the new stack will be:

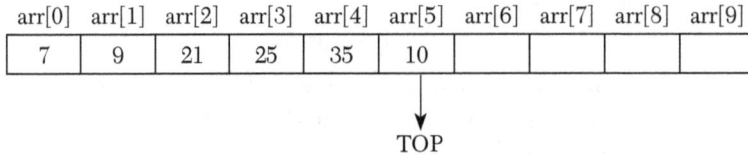

arr[0]	arr[1]	arr[2]	arr[3]	arr[4]	arr[5]	arr[6]	arr[7]	arr[8]	arr[9]
7	9	21	25	35	10				

TOP

Figure 7.3. Stack after inserting a new element.

Algorithm for a push operation in a stack

```
Step 1: START
Step 2: IF TOP = MAX - 1
            Print OVERFLOW ERROR
        Go to Step 5
        [End of If]
Step 3: Set TOP = TOP + 1
Step 4: Set STACK[TOP] = ITEM
Step 5: EXIT
```

In the previous algorithm, first we check for the overflow condition. In Step 3, TOP is incremented so that it points to the next location. Finally, the new element is inserted in the stack at the position pointed to by TOP.

2. POP

The pop operation is the process of removing elements from a stack. However, before deleting an element from a stack, we must always check for the underflow condition, which occurs when we try to delete an element from a stack which is already empty. An underflow condition can be checked as follows, if TOP = NULL. Hence, if the underflow condition is true, then an underflow message is displayed on the screen; otherwise, the element is deleted from the stack.

For example: Let us take a stack which has five elements in it. Suppose we want to delete an element, 35, from the stack; then TOP will be decremented by 1. Thus, the element is deleted from the position pointed to by TOP. Now, let us see how the pop operation occurs in the stack in the following figure:

After deleting 35 from it, the new stack will be:

arr[0]	arr[1]	arr[2]	arr[3]	arr[4]	arr[5]	arr[6]	arr[7]	arr[8]	arr[9]
7	9	21	25	35					

TOP

arr[0]	arr[1]	arr[2]	arr[3]	arr[4]	arr[5]	arr[6]	arr[7]	arr[8]	arr[9]
7	9	21	25						

TOP

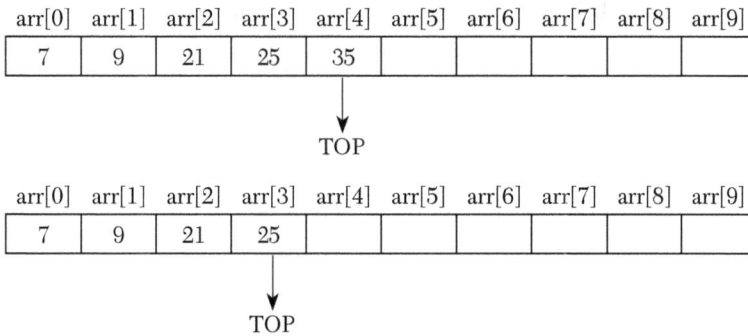

Figure 7.4. Stack after deleting an element.

Algorithm for the pop operation in a stack

```
Step 1: START
Step 2: IF TOP = NULL
            Print UNDERFLOW ERROR
            Go to Step 5
        [End of If]
Step 3: Set ITEM = STACK[TOP]
Step 4: Set TOP = TOP - 1
Step 5: EXIT
```

In the previous algorithm, first we check for the underflow condition, that is, whether the stack is empty or not. If the stack is empty, then no deletion takes place; otherwise, TOP is decremented to the previous position in the stack. Finally, the element is deleted from the stack.

3. PEEK

Peek is an operation that returns the value of the topmost element of the stack. It does so without deleting the topmost element of the array. However, the peek operation first checks for the underflow condition. An underflow condition can be checked as follows, if TOP = NULL. Hence, if the underflow condition is true, then an underflow message is displayed on the screen; otherwise, the value of the element is returned.

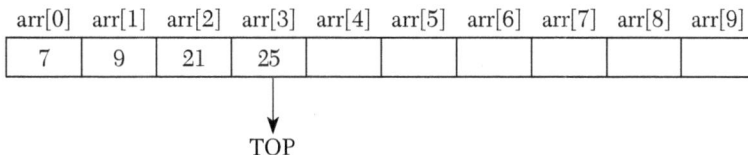

arr[0]	arr[1]	arr[2]	arr[3]	arr[4]	arr[5]	arr[6]	arr[7]	arr[8]	arr[9]
7	9	21	25						

TOP

Figure 7.5. Stack returning the topmost value.

Algorithm for the peek operation in a stack

```
Step 1: START
Step 2: IF TOP = NULL
            Print UNDERFLOW ERROR
            Go to Step 4
        [End of If]
Step 3: Return STACK[TOP]
Step 4: EXIT
```

// Write a menu-driven program for stacks, performing all the operations.

```java
public class Stack {
    static int MAX = 10;
    private int[] data = new int[MAX];
    private int top;
    public void push(int item) throws Exception {
        if (top == MAX) {
            throw new Exception("Stack is Full");
        }
        top++;
        data[top] = item;
    }
    public int pop() throws Exception {
        if (top == 0) {
            throw new Exception("Stack is Empty");
        }
        int temp = peek();
        data[top] = 0;
        top--;
        return temp;
    }
    public int peek() throws Exception {
        if (top == 0) {
            throw new Exception("Stack is Empty");
        }
        return data[top];
    }
    public void display() throws Exception {
        if (top == 0) {
            throw new Exception("Stack is Empty");
        }
        for (int i = top; i > 0; i--) {
            System.out.print(data[i] + "->");
        }
```

```
            System.out.println("NULL");
        }
}
//CLIENT CLASS
import java.util.Scanner;
public class StackClient {
    public static void main(String[] args) {
        Scanner scn = new Scanner(System.in);
        Stack s = new Stack();
        boolean flag = true;
        try {
            while (flag) {
                System.out.println("\n***MENU***");
                System.out.println("1. Push");
                System.out.println("2. Pop");
                System.out.println("3. Display");
                System.out.println("4. Exit");
                System.out.println("Enter your choice: ");
                int choice = scn.nextInt();
                int item = 0;
                switch (choice) {
                case 1:
                    System.out.println("Enter value of node: ");
                    item = scn.nextInt();
                    s.push(item);
                    System.out.println(item + " inserted
                                        successfully");
                    break;
                case 2:
                    System.out.println(s.pop() + " deleted
                                        successfully");
                    break;
                case 3:
                    s.display();
                    break;
                case 4:
                    flag = false;
                    System.out.println("Terminated.....");
                    break;
                default:
                    System.out.println("Wrong choice");
                    break;
                }
```

```
        }
    } catch (Exception e) {
        System.out.println(e.getMessage());
    }
  }
}
```

The output of the program is shown as:

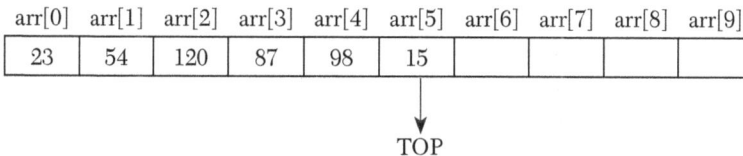

7.5   Implementation of Stacks

Stacks can be represented by two data structures:

1. Representation of stacks using arrays.

2. Representation of stacks using a linked list.

Now, let us discuss both of them in detail.

7.5.1   Implementation of Stacks Using Arrays

Stacks can be easily implemented using arrays. Initially, the TOP of the stack points at the first position or location of the array. As we insert new elements into the stack, the TOP keeps on incrementing, always pointing to the position where the next element will be inserted. The representation of a stack using an array is shown as follows:

arr[0]	arr[1]	arr[2]	arr[3]	arr[4]	arr[5]	arr[6]	arr[7]	arr[8]	arr[9]
23	54	120	87	98	15				

TOP

Figure 7.6. Array representation of a stack.

7.5.2   Implementation of Stacks Using Linked Lists

We have already studied how a stack is implemented using an array. Now let us discuss the same using linked lists. We already know that in linked lists, dynamic memory allocation takes place; that is, the memory is allocated at runtime. But in the case of arrays, memory is allocated at the start of the program. If we are aware of the maximum size of the stack in advance, then implementation of stacks using arrays will be efficient. But if the size is not known in advance, then we will use the concept of a linked list, in which dynamic memory allocation takes place. As we all know a linked list has two parts; the first part contains the information of the node, and the second part stores the address of the next element in the linked list. Similarly, we can also implement a linked stack. Now, the START in the linked list will become the TOP in a linked stack. All insertion and deletion operations will be done at the node pointed to by TOP only.

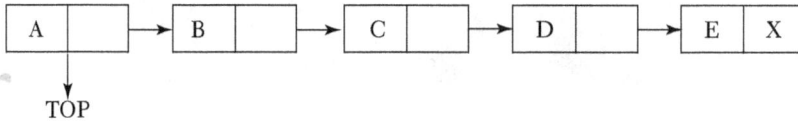

Figure 7.7. Linked representation of a stack.

7.5.2.1   Push Operation in Linked Stacks

The push operation is the process of adding new elements in the already existing stack. The new elements in the stack will always be inserted at the topmost position of the stack. Initially, *we will check whether TOP = NULL.* If the condition is true, then the stack is empty; otherwise, the new memory is allocated for the new node. We will understand it further with the help of an algorithm:

Algorithm for inserting a new element in a linked stack

```
Step 1: START
Step 2: Set NEW NODE.INFO = VAL
        IF TOP = NULL
           Set NEW NODE.NEXT = NULL
           Set TOP  = NEW NODE
        ELSE
           Set NEW NODE.NEXT = TOP
           Set TOP = NEW NODE
        [End of If]
Step 3: EXIT
```

For example: Consider a linked stack with four elements; a new element is to be inserted in the stack.

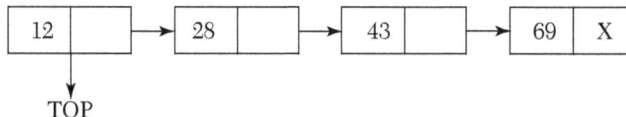

Figure 7.8. Linked stack before insertion.

After inserting the new element in the stack, the updated stack becomes as shown in the following figure:

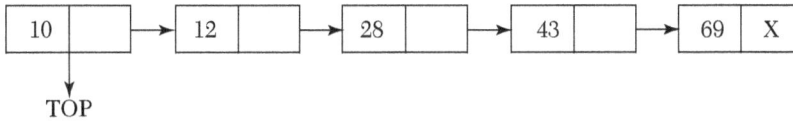

Figure 7.9. Linked stack after inserting a new node.

7.5.2.2 Pop Operation in Linked Stacks

The pop operation is the process of removing elements from an already existing stack. The elements from the stack will always be deleted from the node pointed to by TOP. Initially, we will check whether TOP = NULL. If the condition is true, then the stack is empty, which means we cannot delete any elements from it. Therefore, in that case an underflow error message is displayed on the screen. We will understand it further with the help of an algorithm:

Algorithm for deleting an element from a linked stack

```
Step 1: START
Step 2: IF TOP = NULL
        Print UNDERFLOW ERROR
        [End of If]
Step 3: Set TEMP = TOP
Step 4: Set TOP = TOP.NEXT
Step 5: EXIT
```

For example: Consider a linked stack with five elements; an element is to be deleted from the stack.

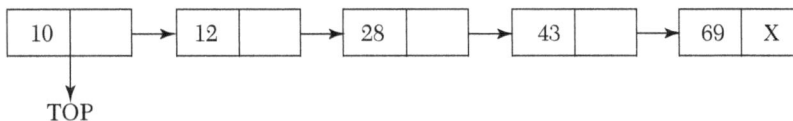

Figure 7.10. Linked stack before deletion.

After deleting an element from the stack, the updated stack becomes as shown in the following figure:

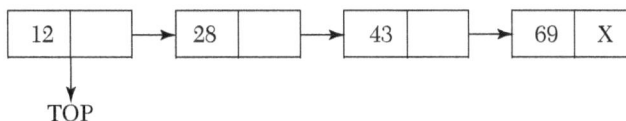

Figure 7.11. Linked stack after deleting the topmost node/element.

```
// Write a menu-driven program implementing a linked stack performing
push and pop operations.
public class StackUsingLinkedList {
    private class Node {
        int data;
        Node next;
    }

        Node top;
        int size;
    private Node create_new_node(int item) throws Exception {
        Node node = new Node();
        if (node == null) {
            throw new Exception("Memory not allocated");
        } else {
            node.data = item;
            node.next = null;
            return node;
        }
    }
    public void push(int item) throws Exception {
        Node nn = create_new_node(item);
        if (size == 0) {
            top = nn;
        } else {
            nn.next = top;
            top = nn;
        }
        size++;
    }
    public int pop() throws Exception {
        if (size == 0) {
            throw new Exception("Stack is Empty");
        }
        int val = peek();
        top = top.next;
        return val;
    }
    public int peek() throws Exception {
        if (size == 0) {
            throw new Exception("Stack is Empty");
        }
        return top.data;
    }
```

```java
    public void display() throws Exception {
        if (size == 0) {
            throw new Exception("Stack is Empty");
        }
        Node temp = top;
        while (temp != null) {
            System.out.print(temp.data + " -> ");
            temp = temp.next;
        }
    System.out.println("NULL");
    }
}
//CLIENT CLASS
import java.util.Scanner;
public class StackUsingLLClient {
    public static void main(String[] args) {
        Scanner scn = new Scanner(System.in);
        StackUsingLinkedList s = new StackUsingLinkedList();
        boolean flag = true;
        try {
            while (flag) {
                System.out.println("\n***MENU***");
                System.out.println("1. Push");
                System.out.println("2. Pop");
                System.out.println("3. Display");
                System.out.println("4. Exit");
                System.out.println("Enter your choice: ");
                int choice = scn.nextInt();
                int item = 0;
                switch (choice) {
                case 1:
                    System.out.println("Enter value of node: ");
                    item = scn.nextInt();
                    s.push(item);
                    System.out.println(item + " inserted
                                        successfully");
                    break;
                case 2:
                    System.out.println(s.pop() + " deleted
                                        successfully");
                    break;
                case 3:
                    s.display();
```

```
                    break;
            case 4:
                    flag = false;
                    System.out.println("Terminated.....");
                    break;
            default:
                    System.out.println("Wrong choice");
                    break;
            }
        }
    } catch (Exception e) {
        System.out.println(e.getMessage());
    }
    }
}
```

The output of the program is shown as:

7.6 Applications of Stacks

In this section, we will discuss various applications of stacks. The topics that will be covered in this section are the following:

- Polish and Reverse Polish Notations
- Conversion from Infix Expression to Postfix Expression
- Conversion from Infix Expression to Prefix Expression
- Evaluation of Postfix Expression
- Evaluation of Prefix Expression
- Parenthesis Balancing

Now, let us understand each one of them in detail.

7.6.1 Polish and Reverse Polish Notations
(a) Polish Notations

Polish notation refers to a notation in which the operator is placed before the operands. Polish notation was named after Polish mathematician Jan Lukasiewicz. We can also say that transforming an expression into a form is called a Polish notation. An algebraic expression can be represented in three forms. All these forms refer to the relative position of operators with respect to the operands.

1. **Prefix Form**: In an expression, if the operator is placed before the operands, that is, +XY, then it is said to be in prefix form.

2. **Infix Form**: In an expression, if the operator is placed in the middle of the operands, that is, X + Y, then it is said to be in infix form.

3. **Postfix Form**: In an expression, if the operator is placed after the operands, that is, XY+, then it is said to be in postfix form.

(b) Reverse Polish Notation

This notation frequently refers to the postfix notation or suffix notation. It refers to the notation in which the operator is placed after its two operands, that is, XY + AF BC∗.

(c) Need for Polish and Reverse Polish Notation

It is comparatively easy for a computer system to evaluate an expression in Polish notation; the system need not check for priority-wise execution of various operators (like the BODMAS (Bracket Of Division, Multiplication, Addition, Subtraction) Rule), as all the operators in prefix or postfix expressions will automatically occur in their order of priority.

7.6.2 Conversion from Infix Expression to Postfix Expression

In any expression, we observe that there are two types of parts/ components clubbed together. They are operands and operators. The operators are the ones that indicate the operation to be carried out, and the operands are ones on which the operators operate. Operators have their priority of execution. For simplicity of the algorithm, we will use only addition (+), subtraction(-), modulus (%), multiplication (∗), and division (/) operators. The precedence of these operators is given as follows:

∗, ^, /, % (Higher priority)   +, - (Lower priority)

The order of evaluation of these operators can be changed by using parentheses. For example, an expression X ∗ Y + Z can be solved, as first X ∗ Y will be done and then the result is added to Z. But if the same expression is written with parentheses as X ∗ (Y + Z), now Y + Z will be evaluated first, and then the result is multiplied by X.

We can convert an infix expression to a postfix expression using a stack. First, we start to scan the expression from the left side to the right side. In an expression, there may be some operators, operands, and parentheses.

Hence, we have to keep in mind some of the basic rules, which are as follows:

- Each time we encounter an operand, it is added directly to the postfix expression.

- Each time we get an operator, we should always check the top of the stack to check the priority of the operators.

- If the operator at the top of the stack has higher precedence or the same precedence as that of the current operator, then it is repeatedly popped out from the stack and added to the postfix expression. Otherwise, it is pushed into the stack.

- Each time when an opening parenthesis is encountered, it is directly pushed into the stack and, similarly, if a closing parenthesis is encountered, we will repeatedly pop it out from the stack and add the operators in the postfix expression. Also, the opening parenthesis is deleted from the stack.

Now, let us understand it with the help of an algorithm in which the first step is to push a left parenthesis in the stack and also add a closing parenthesis at the end of the infix expression. The algorithm is repeated until the stack becomes empty.

Algorithm to convert an infix expression into a postfix expression

```
Step 1: START
Step 2: Add ")" parenthesis to the end of infix expression.
Step 3: Push' 'parenthesis on the stack.
Step 4: Repeat the steps until each character in the infix
        expression is scanned.
        (a) IF "(" parenthesis is found, push it onto the stack.
        (b) If an operand is encountered, add it to the postfix
            expression.
        (c) IF ")" parenthesis is found, then follow these steps:
            - Continually pop from the stack and add it to the postfix
              expression until a "(" parenthesis is encountered.
            - Eliminate the "(" parenthesis.
        (d) If an operator is found, then follow these steps:
            - Continually pop from the stack and add it to the post-
              fix expression which has same or high precedence than
              the current operator.
            - Push the current operator to the stack.
```

Step 5: Continually pop from the stack to the postfix expression until the stack becomes empty.

Step 6: EXIT

For example: Convert the following infix expression into a postfix expression.

(a) $(A + B) * C / D$

(b) $[((A + B) * (C - D)) + (F - G)]$

Solution:

(a)

Character	Stack	Expression
(	(	
A	(	A
+	(+	A
B	(+	AB
)		AB+
*	*	AB+
C	*	AB+C
/	/	AB+C*
D		AB+C*D/
		Answer = AB+C*D/

(b)

Character	Stack	Expression
[	[	
(	[(	
(	[((	
A	[((	A
+	[((+	A
B	[((+	AB
)	[(	AB+
*	[(*	AB+
(	[(*(	AB+
C	[(*(	AB+C
-	[(*(-	AB+C
D	[(*(-	AB+CD
)	[(*	AB+CD-

Character	Stack	Expression
)	[	AB+CD–*
+	[+	AB+CD–*
(	[+(	AB+CD–*
F	[+(	AB+CD–*F
–	[+(–	AB+CD–*F
G	[+(–	AB+CD–*FG
)	[+	AB+CD–*FG-
]		AB+CD–*FG–+
		Answer = AB+CD–*FG–+

// Write a program to convert an infix expression to a postfix expression.

```java
import java.util.Scanner;
public class InfixToPostfixConversion {
    static int COUNT;
    public static String convert(String str) {
        String ans = "";
        char[] stack = new char[str.length()];
        for (int i = 0; i < str.length(); i++) {
            char c = str.charAt(i);
            if (priority(c) > 0) {
                while (isEmpty(stack) == false &&
                        priority(peek(stack)) >= priority(c)) {
                    ans += pop(stack);
                }
                push(stack, c);
            } else if (c == ')') {
                char x = pop(stack);
                while (x != '(') {
                    ans += x;
                    x = pop(stack);
                }
            } else if (c == '(') {
                push(stack, c);
            } else {
                ans += c;
            }
        }
        for (int i = 0; i <= size(stack); i++) {
            ans += pop(stack);
        }
        return ans;
    }
}
```

```java
public static int size(char[] stack) {
    return COUNT;
}
public static void push(char[] stack, char c) {
    COUNT++;
    stack[COUNT - 1] = c;
}
public static char pop(char[] stack) {
    char ch = stack[COUNT - 1];
    COUNT--;
    return (ch);
}
public static char peek(char[] stack) {
    if (size(stack) == 0)
        return 0;
    else
        return stack[size(stack) - 1];
}
public static boolean isEmpty(char[] stack) {
    if (stack[0] == 0)
        return true;
    else
        return false;
}
public static int priority(char c) {
    switch (c) {
    case '+':
        return 1;
    case '-':
        return 1;
    case '*':
        return 2;
    case '/':
        return 2;
    case '%':
        return 2;
    case '^':
        return 3;
    default:
        return -1;
    }
}
public static void main(String[] args) {
    Scanner scn = new Scanner(System.in);
    System.out.println("Enter your Infix Expression");
```

```
        String exp = scn.next();
        System.out.println("Infix Expression: " + exp);
        System.out.println("Postfix Expression: " + convert(exp));
    }
}
```

The output of the program is shown as:

Frequently Asked Questions

2. Convert the following infix expression into a postfix expression.
(A + B) ^ C – (D * E) / F
Ans:

Character	Stack	Expression
(	(	
A	(	A
+	(+	A
B	(+	AB
)		AB+
^	^	AB+
C	^	AB+C
-	-	AB+C^
(	-(	AB+C^
D	-(	AB+C^D
*	-(*	AB+C^D
E	-(*	AB+C^DE
)	-	AB+C^DE*
/	-/	AB+C^DE*
F	-/	AB+C^DE*
		Answer = AB+C^DE*F/-

7.6.3 Conversion from Infix Expression to Prefix Expression

We can convert an infix expression to its equivalent prefix expression with the help of the following algorithm.

Algorithm to convert an infix expression into a prefix expression

Step 1: START

Step 2: Reverse the infix expression. Also, interchange left and right parenthesis on reversing the infix expression.

Step 3: Obtain the postfix expression of the reversed infix expression.

Step 4: Reverse the postfix expression so obtained in Step 3. Finally, the expression is converted into prefix expression.

Step 5: EXIT

For example: Convert the following infix expression into a prefix expression.

(a) (X - Y) / (A + B)

(b) (X – Y / Z) * (A / B – C)

Solution:

(a) After reversing the given infix expression((B + A) / Y – X)

Find the postfix expression of (B + A) / (Y – X)

Character	Stack	Expression
(	(	
(	((	
B	((	B
+	((+	B
A	((+	BA
)	(	BA+
/	(/	BA+
Y	(/	BA+Y
-	(-	BA+Y/
X	(-	BA+Y/X
)		BA+Y/X-
		BA+Y/X-

Now, reverse the postfix expression so obtained, that is, X/Y+AB

Hence, the prefix expression is **–X/Y+AB**

(b) After reversing the given infix expression (C – B / A) ∗ (Z / Y – X)

Find the postfix expression of (C – B / A) ∗ (Z / Y – X)

Character	Stack	Expression
(	(	
C	(	C
-	(-	C
B	(-	CB
/	(-/	CB
A	(-/	CBA
)		CBA/-
∗	∗	CBA/-
(	∗(	CBA/-
Z	∗(	CBA/-Z
/	∗(/	CBA/-Z
Y	∗(/	CBA/-ZY
-	∗(-	CBA/-ZY/
X	∗(-	CBA/-ZY/X
)	∗	CBA/-ZY/X-
		CBA/-ZY/X-∗

Now, reverse the postfix expression so obtained, that is, ∗–X/ZY-/ABC

Hence, the prefix expression is **∗–X/ZY-/ABC**

// Write a program to convert an infix expression to a prefix expression.

```java
import java.util.Scanner;
public class InfixToPrefixConversion {
    static int COUNT;
    public static String convert(String inp) {
        String str = reverse(inp);
        str = reverseBrackets(str);
        String ans = "";
        char[] stack = new char[str.length()];
        for (int i = 0; i < str.length(); i++) {
            char c = str.charAt(i);
            if (priority(c) > 0) {
                while (isEmpty(stack) == false &&
                        priority(peek(stack)) >= priority(c)) {
                    ans += pop(stack);
                }
```

```
                push(stack, c);
            } else if (c == ')' && COUNT > 0) {
                char x = pop(stack);
                while (x != '(') {
                    ans += x;
                    x = pop(stack);
                }
            } else if (c == '(') {
                push(stack, c);
            } else {
                ans += c;
            }
        }
        for (int i = 0; i <= size(stack); i++) {
            ans += pop(stack);
        }
        String result = reverse(ans);
        return result;
    }
    public static int size(char[] stack) {
        return COUNT;
    }
    public static void push(char[] stack, char c) {
        COUNT++;
        stack[COUNT - 1] = c;
    }
    public static char pop(char[] stack) {
        char ch = stack[COUNT - 1];
        COUNT--;
        return (ch);
    }
    public static char peek(char[] stack) {
        if (size(stack) == 0)
            return 0;
        else
            return stack[size(stack) - 1];
    }
    public static boolean isEmpty(char[] stack) {
        if (COUNT == 0)
            return true;
        else
            return false;
    }
    public static int priority(char c) {
```

```
    switch (c) {
    case '+':
        return 1;
    case '-':
        return 1;
    case '*':
        return 2;
    case '/':
        return 2;
    case '%':
        return 2;
    case '^':
        return 3;
    default:
        return -1;
    }
}
public static String reverse(String inp) {
    String result = "";
    for (int i = inp.length() - 1; i >= 0; i--) {
        result += inp.charAt(i);
    }
        return result;
}
public static String reverseBrackets(String str) {
    String rev = "";
    for (int i = 0; i < str.length(); i++) {
        char c = str.charAt(i);
        if (c == '(')
            rev += ")";
        else if (c == ')')
            rev += "(";
        else
            rev += c;
    }
    return rev;
}
public static void main(String[] args) {
    Scanner scn = new Scanner(System.in);
    System.out.println("Enter your Infix Expression");
    String exp = scn.next();
    System.out.println("Infix Expression: " + exp);
    System.out.println("Prefix Expression: " + convert(exp));
}
```

}

The output of the program is shown as:

7.6.4 Evaluation of a Postfix Expression

With the help of stacks, any postfix expression can easily be evaluated. Every character in the postfix expression is scanned from left to right. The steps involved in evaluating a postfix expression are given in the algorithm.

Algorithm for evaluating a postfix expression

```
Step 1: START
Step 2: IF an operand is encountered, push it onto the stack.
Step 3: IF an operator "op1" is encountered, then follow these
        steps:
        (a) Pop the two topmost elements from the stack, where X
            is the topmost element and Y is the next top element
            below X.
        (b) Evaluate X op1 Y.
        (c) Push the result onto the stack.
Step 4: Set the result equal to the topmost element of the stack.
Step 5: EXIT
```

For example: Evaluate the following postfix expressions.

(a) 2 3 4 + * 5 6 7 8 + * + +

(b) T F T F AND F F F XOR OR AND T XOR AND OR

Solution

(a)

Character	Stack	Operation
2	2	PUSH 2
3	2, 3	PUSH 3

Character	Stack	Operation
4	2, 3, 4	PUSH 4
+	2, 7	POP 4, 3 ADD(4 + 3 = 7) PUSH 7
*	14	POP 7, 2 MUL(7 * 2 = 14) PUSH 14
5	14, 5	PUSH 5
6	14, 5, 6	PUSH 6
7	14, 5, 6, 7	PUSH 7
8	14, 5, 6, 7, 8	PUSH 8
+	14, 5, 6, 15	POP 8, 7 ADD(8 + 7 = 15) PUSH 15
*	14, 5, 90	POP 15, 6 MUL(15 * 6 = 90) PUSH 90
+	14, 95	POP 90, 5 ADD(90 + 5 = 95) PUSH 95
+	109	POP 95, 14 ADD(95 + 14 = 109) PUSH 109
	Answer = 109	

(b)

Character	Stack	Operation
T	T	PUSH T
F	T, F	PUSH F
T	T, F, T	PUSH T
F	T, F, T, F	PUSH F
AND	T, F, F	POP F, T AND(F AND T = F) PUSH F
F	T, F, F, F	PUSH F
F	T, F, F, F, F	PUSH F
F	T, F, F, F, F, F	PUSH F
XOR	T, F, F, F, T	POP F, F XOR(F XOR F = T) PUSH T

Character	Stack	Operation
OR	T, F, F, T	POP T, F OR(T OR F = T) PUSH T
AND	T, F, F	POP T, F AND(T AND F = F) PUSH F
T	T, F, F, T	PUSH T
XOR	T, F, F	POP T, F XOR(T XOR F = F) PUSH F
AND	T, F	POP F, F AND(F AND F = F) PUSH F
OR	T	POP F, T OR(F OR T = T) PUSH T
	Answer = T	

// Write a program for evaluation of a postfix expression.

```java
import java.util.Scanner;
public class PostfixEvaluation {
    static int COUNT;
    static int evaluate(String exp) {
        int[] stack = new int[exp.length()];
        for (int i = 0; i < exp.length(); i++) {
            char c = exp.charAt(i);
            if (Character.isDigit(c))
                push(stack, c - '0');
            else {
                int val1 = pop(stack);
                int val2 = pop(stack);
                switch (c) {
                case '+':
                    push(stack, val2 + val1);
                    break;
                case '-':
                    push(stack, val2 - val1);
                    break;
                case '/':
                    push(stack, val2 / val1);
                    break;
```

```
                        case '*':
                            push(stack, val2 * val1);
                            break;
                    }
                }
            }
        return pop(stack);
    }
    public static int size(int[] stack) {
        return COUNT;
    }
    public static void push(int[] stack, int c) {
        COUNT++;
        stack[COUNT - 1] = c;
    }
    public static int pop(int[] stack) {
        int ch = stack[COUNT - 1];
        COUNT--;
        return (ch);
    }
    public static int peek(int[] stack) {
        if (size(stack) == 0)
            return 0;
        else
            return stack[size(stack) - 1];
    }
    public static boolean isEmpty(int[] stack) {
        if (COUNT == 0)
            return true;
        else
            return false;
    }
    public static void main(String[] args) {
        Scanner scn = new Scanner(System.in);
        System.out.println("Enter your Postfix Expression");
        String exp = scn.next();
        System.out.println("Post Evaluation: " + evaluate(exp));
    }
}
```

The output of the program is shown as:

Frequently Asked Questions

Evaluate the given postfix expression.

2 3 4 * 6 / +

Ans:

Character	Stack
2	2
3	2, 3
4	2, 3, 4
*	2, 12
6	2, 12, 6
/	2, 2
+	4
	Answer = 4

7.6.5 Evaluation of a Prefix Expression

There are a variety of techniques for evaluating a prefix expression. But the simplest of all the techniques is explained in the following algorithm.

Algorithm for evaluating a prefix expression

```
Step 1: START
Step 2: Accept the prefix expression.
Step 3: Repeat the steps 4 to 6 until all the characters have been
        scanned.
Step 4: The prefix expression is scanned from the right.
```

Step 5: IF an operand is encountered, push it onto the stack.
Step 6: IF an operator is encountered, then follow these steps:
      (a) Pop two elements from the operand stack.
      (b) Apply the operator on the popped operands.
      (c) Push the result onto the stack.
Step 7: EXIT

For example: Evaluate the given prefix expressions.

(a) + - 4 6 * 9 /10 50

(b) + * * + 2 3 4 5 + 6 7

Solution:

(a)

Character	Stack	Operation
50	50	PUSH 50
10	50, 10	PUSH 10
/	5	POP 10, 50 DIV(50 / 10 = 5) PUSH 5
9	5, 9	PUSH 9
*	45	POP 9, 5 MUL(5 * 9 = 45) PUSH 45
6	45, 6	PUSH 6
4	45, 6, 4	PUSH 4
-	45, 2	POP 4, 6 SUB(6 – 4 = 2) PUSH 2
+	47	POP 2, 45 ADD(45 + 2 = 47) PUSH 47
	Answer = 47	

(a)

Character	Stack	Operation
7	7	PUSH 7
6	7, 6	PUSH 6
+	13	POP 6, 7 ADD(7 + 6 = 13) PUSH 13
5	13, 5	PUSH 5

Character	Stack	Operation
4	13, 5, 4	PUSH 4
3	13, 5, 4, 3	PUSH 3
2	13, 5, 4, 3, 2	PUSH 2
+	13, 5, 4, 5	POP 2, 3 ADD(3 + 2 = 5) PUSH 5
*	13, 5, 20	POP 5, 4 MUL(4 * 5 = 20) PUSH 20
*	13, 100	POP 20, 5 MUL(5 * 20 = 100) PUSH 100
+	113	POP 100, 13 ADD(13 + 100 = 113) PUSH 113
	Answer = 113	

// Write a program for evaluation of a prefix expression.

```java
import java.util.Scanner;
public class PrefixEvaluation {
    static int COUNT;
    static int evaluate(String exp) {
        int[] stack = new int[exp.length()];
        for (int i = 0; i < exp.length(); i++) {
            char c = exp.charAt(i);
            if (Character.isDigit(c))
                push(stack, c - '0');
            else {
                int val2 = pop(stack);
                int val1 = pop(stack);
                switch (c) {
                case '+':
                    push(stack, val1 + val2);
                    break;
                case '-':
                    push(stack, val1 - val2);
                    break;
                case '/':
                    push(stack, val1 / val2);
                    break;
```

```
                case '*':
                    push(stack, val1 * val2);
                    break;
            }
        }
    }
    return pop(stack);
}
public static int size(int[] stack) {
    return COUNT;
}
public static void push(int[] stack, int c) {
    COUNT++;
    stack[COUNT - 1] = c;
}
public static int pop(int[] stack) {
    int ch = stack[COUNT - 1];
    COUNT--;
    return (ch);
}
public static int peek(int[] stack) {
    if (size(stack) == 0)
        return 0;
    else
        return stack[size(stack) - 1];
}
public static boolean isEmpty(int[] stack) {
    if (COUNT == 0)
        return true;
    else
        return false;
}
public static String reverse(String inp) {
    String result = "";
    for (int i = inp.length() - 1; i >= 0; i--) {
        result += inp.charAt(i);
    }
    return result;
}
public static void main(String[] args) {
    Scanner scn = new Scanner(System.in);
    System.out.println("Enter your Prefix Expression");
    String exp = scn.next();
```

```
        exp = reverse(exp);
        System.out.println("Pre Evaluation: " + evaluate(exp));
    }
}
```

The output of the program is shown as:

7.6.6 Parenthesis Balancing

Stacks can be used to check the validity of parentheses in any arithmetic or algebraic expression. We are already aware that in a valid expression, the parentheses or the brackets occur in pairs; that is, if a parenthesis is opening, then it must be closed in an expression. Otherwise, the expression would be invalid. For example, (X + Y − Z is invalid. But (X + Y − Z) looks like a valid expression. Hence, there are some key points which are to be kept in mind:

- Each time a "(" parenthesis is encountered, it should be pushed onto the stack.

- Each time a ")" parenthesis is encountered, the stack is examined.

- If the stack is already an empty stack, then the ")" parenthesis does not have a "(" parenthesis, and hence the expression is invalid.

- If the stack is not empty, then we will pop the stack and check whether the popped element corresponds to the ")" parenthesis.

- When we reach the end of the stack, the stack must be empty. Otherwise, one or more "(" parenthesis does not have a corresponding ")" parenthesis and, therefore, the expression will become invalid.

For example: Check whether the following given expressions are valid or not.

(a) ((A − B) * Y

(b) [(A + B) − {X + Y} * [C − D]]

Solution:

	Symbol	Stack
1.	(	(
2.	(	(, (
3.	A	(, (
4.	-	(, (
5.	B	(, (
6.	)	(
7.	*	(
8.	Y	(
9.		(

Answer: As the stack is not empty, the expression is not a valid expression.

	Symbol	Stack
1.	[	[
2.	(	[, (
3.	A	[, (
4.	+	[, (
5.	B	[, (
6.	)	[
7.	-	[
8.	{	[, {
9.	X	[, {
10.	+	[, {
11.	Y	[, {
12.	}	[
13.	*	[
14.	[	[, [
15.	C	[, [
16.	-	[, [
17.	D	[, [
18.	]	[
19.	]	

Answer: As the stack is empty, the given expression is a valid expression.

// Write a program to implement parenthesis balancing.

```java
import java.util.Scanner;
public class BalancedParenthesis {
    static class stack {
        int top = -1;
        char arr[] = new char[30];
        public void push(char x) {
            if (top == arr.length - 1)
                System.out.println("Stack full");
            else
                arr[++top] = x;
        }
        public char pop() {
            if (top == -1)
                System.out.println("Underflow error");
            else
                return arr[top--];
            return 0;
        }
        boolean isEmpty() {
            if (top == -1)
                return true;
            else
                return false;
        }
    }
    static boolean isPair(char c1, char c2) {
        if (c1 == '(' && c2 == ')')
            return true;
        else if (c1 == '{' && c2 == '}')
            return true;
        else if (c1 == '[' && c2 == ']')
            return true;
        else
            return false;
    }
    static boolean isBalanced(char exp[]) {
        stack s = new stack();
        for (int i = 0; i < exp.length; i++) {
            if (exp[i] == '{' || exp[i] == '(' || exp[i] == '[')
                s.push(exp[i]);
            if (exp[i] == '}' || exp[i] == ')' || exp[i] == ']') {
                if (s.isEmpty()) {
```

```
                    return false;
                } else if (!isPair(s.pop(), exp[i])) {
                    return false;
                }
            }
        }
        if (s.isEmpty())
            return true;
        else {
            return false;
        }
    }
    public static void main(String[] args) {
        Scanner scn = new Scanner(System.in);
        System.out.println("Enter the expression:");
        String input = scn.next();
        char[] exp = new char[input.length()];
        for (int i = 0; i < exp.length; i++) {
            exp[i]=input.charAt(i);
        }
        if (isBalanced(exp))
            System.out.println("Balanced ");
        else
            System.out.println("Not Balanced ");
    }
}
```

The output of the program is shown as:

7.7 Summary

- A stack is a linear collection of data elements in which the element inserted last will be the element taken out first (i.e., a stack is a LIFO data structure). The stack is a linear data structure, in which the insertion as well as the deletion of an element is done only from the end called TOP.

- In computer memory, stacks can be implemented by using either arrays or linked lists.

- The overflow condition occurs when we try to insert the elements in the stack, but the stack is already full.

- The underflow condition occurs when we try to remove the elements from the stack, but the stack is already empty.

- The three basic operations that can be performed on the stacks are push, pop, and peek operations.

- A push operation is the process of adding new elements in the stack.

- A pop operation is the process of removing elements from the stack.

- A peek operation is the process of returning the value of the topmost element of the stack.

- Polish notation refers to a notation in which the operator is placed before the operands.

- Infix, prefix, and postfix notations are three different but equivalent notations of writing algebraic expressions.

7.8 Exercises

7.8.1 Theory Questions

1. What is a stack? Give its real-life example.

2. What do you understand about stack overflow and stack underflow?

3. What is a linked stack, and how it is different from a linear stack?

4. Discuss various operations which can be performed on the stacks.

5. Explain the terms Polish notation and reverse Polish notation.

6. What are the various applications of a stack? Explain in detail.

7. Why is a stack known as a Last-In-First-Out structure?

8. What are different notations to represent an algebraic expression? Which one is mostly used in computers?

9. Explain the concept of linked stacks and also discuss how insertion and deletion takes place in it.

10. Draw the stack structure when the following operations are performed one after another on an empty stack.
   (a) Push 1, 2, 6, 17, 100
   (b) Pop three numbers
   (c) Peek
   (d) Push 50, 23, 198, 500
   (e) Display

11. Convert the following infix expressions to their equivalent postfix expressions.
   (a) A + B + C − D ∗ E / F
   (b) [A − C] + {D ∗ E}
   (c) [X / Y] % (A ∗ B) + (C % D)
   (d) [(A − C + D) % (B − H + G)]
   (e) 18 / 9 ∗ 3 − 4 + 10 / 2

12. Check the validity of the given algebraic expressions.
   (a) ((([A − V − D] + B)
   (b) [(X − {Y ∗ Z})]
   (c) [A + C + E)

13. Convert the following infix expressions to their equivalent prefix expressions.
   (a) 18 / 9 ∗ 3 − 4 + 10 / 2
   (b) X ∗ (Z / Y)
   (c) [(A + B) − (C + D)] ∗ E

14. Evaluate the given postfix expressions.
   (a) 1 2 3 ∗ ∗ 4 5 6 7 + + ∗ ∗
   (b) 12 4 / 4 5 + 2 3 ∗ +

7.8.2 Programming Questions

1. Write a Java program to implement a stack using arrays.

2. Write a program to convert an infix expression to a prefix expression.

3. Write a program to copy the contents from one stack to another.

4. Write a Java program to convert the expression "x + y" into "xy+".

5. Write a program to evaluate a postfix expression.

6. Write a program to evaluate a prefix expression.

7. Write a program to convert "b - c" into "-bc".

8. Write a function that performs a push operation in a linked stack.

7.8.3 Multiple Choice Questions

1. New elements in the stack are always inserted from:
   (a) Frontend
   (b) Top end
   (c) Rear end
   (d) Both (a) and (c)

2. A stack is a _____ data structure.
   (a) FIFO
   (b) LIFO
   (c) FILO
   (d) LILO

3. The overflow condition in the stack exists when:
   (a) TOP = NULL
   (b) TOP = MAX
   (c) TOP = MAX − 1
   (d) None of the above

4. The function that inserts the elements in a stack is called _____.
   (a) Push()
   (b) Peek()
   (c) Pop()
   (d) None of the above

5. Discs piled up one above the other represent a _____.

 (a) Queue

 (b) Stack

 (c) Tree

 (d) Linked List

6. Reverse Polish notation is the other name for a _____.

 (a) Postfix expression

 (b) Prefix expression

 (c) Infix expression

 (d) All of the above

7. Stacks can be represented by:

 (a) Linked List only

 (b) Arrays only

 (c) Both a) and b)

 (d) None of the above

8. If the numbers 10, 45, 13, 50, and 32 are pushed onto a stack, what does pop return?

 (a) 10

 (b) 45

 (c) 50

 (d) 32

9. The postfix representation of the expression $(2 - b) * (a + 10) / (c * 8)$ will be:

 (a) 8 a $*$ c 10 + b 2 - $*$ /

 (b) / 2 a c $*$ + b 10 $*$ 9 −

 (c) 2 b − a 10 + $*$ c 8 $*$ /

 (d) 10 a + $*$ 2 b - / c 8 $*$

CHAPTER 8

TREES

8.1 Introduction

In earlier chapters we learned about various data structures such as arrays, linked lists, stacks, and queues. All these data structures are linear data structures. Although linear data structures are flexible, it is quite difficult to use them to organize data into a hierarchical representation. Hence, to overcome this problem or limitation, we create a new data structure which is called a tree. *A tree is a data structure that is defined as a set of one or more nodes which allows us to associate a parent-child relationship.* In trees, one node is designated as the root node or parent node, and all the remaining nodes can be partitioned into non-empty sets, each of which is a subtree of the root. Unlike natural trees, a tree data structure is upside down, having a root at the top and leaves at the bottom. Also, there is no parent of the root node. A root node can only have child nodes. On the contrary, leaf nodes or leaves are those that have no children. *When there are no nodes in the tree, then the tree is known as a null tree or empty tree.* Trees are widely used in various day-to-day applications. Also, the recursive programming of trees makes the programs optimized and easily understandable. Trees are also used to represent the structure of mathematical formulas. Figure 8.1 represents a tree. In the following tree, A is the root node of the tree. X, Y, and Z are the child nodes of the root node A. They also form the subtrees of the tree. Also, B, C, Y, D, and E are the leaf nodes of the tree, as they have no children.

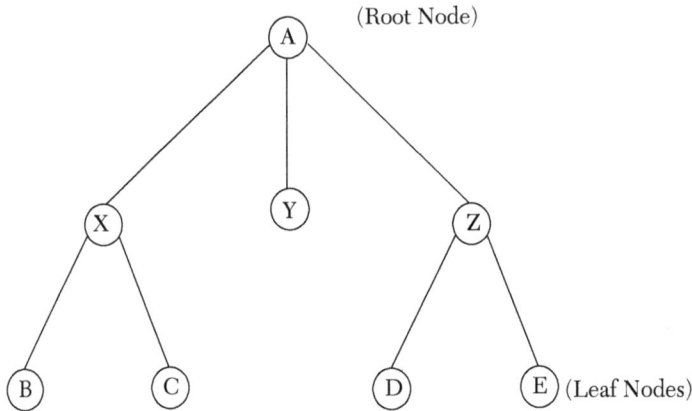

Figure 8.1. A tree.

Practical Application

1. The members of a family can be visualized as a tree in which the root node can be visualized as a grandfather. His two children can be visualized as the child nodes. Then the grandchildren form the left and the right subtrees of the tree.

2. Trees are used to organize information in database systems and also to represent the syntactic structure of the source programs in compilers.

8.2 Definitions

- **Node:** A node is the main component of the tree data structure. It stores the actual data along with the links to the other nodes.

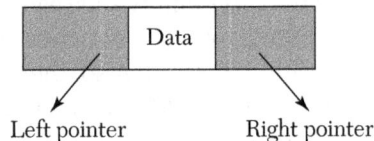

Figure 8.2. Structure of a node.

- **Root:** The root node is the topmost node of the tree. It does not have a parent node. If the root node is empty, then the tree is empty.

- **Parent:** The parent of a node is the immediate predecessor of that node. In the following figure, X is the parent of the Y and Z nodes.

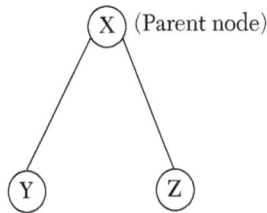

Figure 8.3. Parent node.

- **Child:** The child nodes are the immediate successors of a node. They must have a parent node. A child node placed at the left side is called the left child and, similarly, a child node placed at the right side is called a right child. Y is the left child of X, and Z is the right child of X.

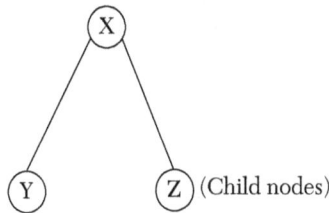

Figure 8.4. Child nodes.

- **Leaf/Terminal nodes:** A leaf node is one which does not have any child nodes.

- **Subtrees:** The nodes B, X, and Y form the left subtree of root A. Similarly, the nodes C and Z form the right subtree of A.

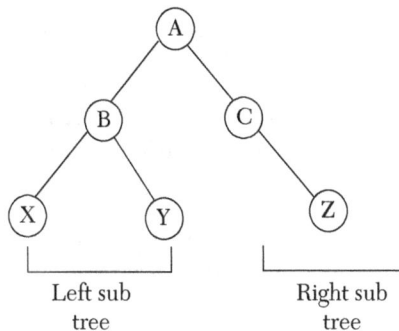

Figure 8.5. Subtrees.

- **Path:** It is a unique sequence of consecutive edges which is required to be followed to reach the destination from a given source. The path from root node A to Y is given as A-B, B-Y.

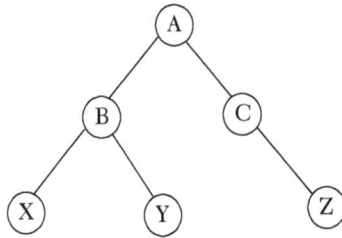

Figure 8.6. Path.

- **Level number of a node:** Every node in the tree is assigned a level number. The root is at level 0, the children of the root node are at level 1, and so on.

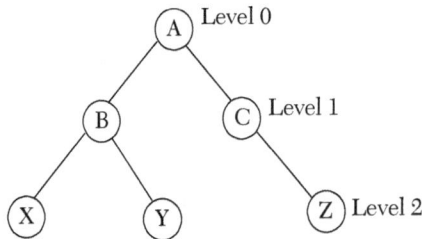

Figure 8.7. Node level numbers.

- **Height:** The height of the tree is the maximum level of the node + 1. The height of a tree containing a single node will be 1. Similarly, the height of an empty tree will be 0.

- **Ancestors:** The ancestors of a node are any predecessor nodes on the path between the root and the destination. There are no ancestors for the root node. The nodes A and B are the ancestors of node X.

- **Descendants:** The descendants of a node are any successor nodes on the path between the given source and the leaf node. There are no descendants of the leaf node. Here, B, X, and Y are the descendants of node A.

- **Siblings:** The child nodes of a given parent node are called siblings. X and Y are the siblings of B in Figure 8.8.

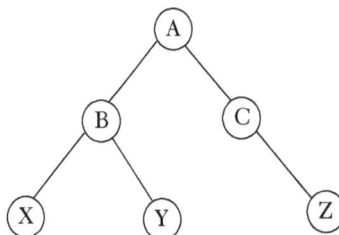

Figure 8.8. Siblings.

- **Degree of a node:** It is equal to the number of children that a node has.

- **Out-degree of a node:** It is equal to the number of edges leaving that node.

- **In-degree of a node:** It is equal to the number of edges arriving at that node.

- **Depth:** It is given as the length of the path from the root node to the destination node.

8.3 Binary Tree

A binary tree is a collection of nodes where each node contains three parts, that is, the left child address, the right child address, and the data item. The left childaddress stores the memory location of the top node of the left subtree and the right child address stores the memory location of the top node of the right subtree. The topmost element of the binary tree is known as a root node. The root stores the memory location of the root node. As the name suggests, *a binary tree can have at most two children, that is, a parent can have zero, one, or at most two children.* Also, if root = NULL, then it means that the tree is empty. Figure 8.9 represents a binary tree.

In the following figure, A represents the root node of the tree. B and C are the children of root node A. Nodes B, D, E, F, and G constitute the left subtree. Similarly, nodes C, H, I, and J constitute the right subtree. Now, nodes G, E, F, I, and J are the terminal/leaf nodes of the binary tree, as they have no children. Hence, node A has two successors B and C. Node B has two successors D and G. Similarly, node D also has two successors E and F. Node G has no successor. Node C has only one successor H. Node H has two successors I and J. Since nodes E, F, G, I, and J have no successors, they are said to have empty subtrees.

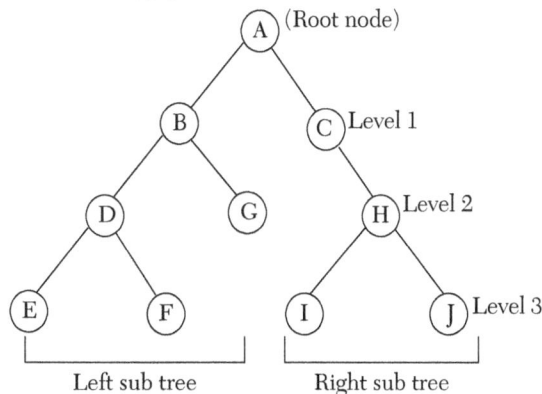

Figure 8.9. A binary tree.

8.3.1 Types of Binary Trees

There are two types of binary trees:

1. **Complete Binary Trees:** A complete binary tree is a type of binary tree which obeys/satisfies two properties:

   (a) First, every level in a complete binary tree except the last one must be completely filled.

   (b) Second, all the nodes in the complete binary tree must appear left as much as possible.

   In a complete binary tree, the number of nodes at level n is 2^n nodes. Also, the total number of nodes in a complete binary tree of depth d is equal to the sum of all nodes present at each level between 0 and d.

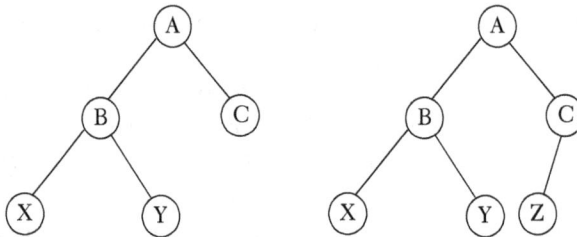

Figure 8.10. Complete binary trees.

2. **Extended Binary Trees:** Extended binary trees are also known as 2T-trees. A binary tree is said to be an extended binary tree if and only if every node in the tree has either zero children or two children. In an extended binary tree, nodes having two children are known as internal nodes. On the contrary, nodes having no children are known as external nodes. In the following figure, the internal nodes are represented by I and the external nodes are represented by E.

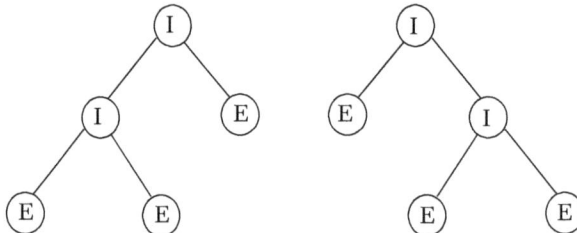

Figure 8.11. Extended binary trees.

8.3.2 Memory Representation of Binary Trees

Binary trees can be represented in a computer's memory in either of the following ways:

1. Array Representation of Binary Trees

2. Linked Representation of Binary Trees

Now, let us discuss both of them in detail.

Array Representation of Binary Trees

A binary tree is represented using an array in the computer's memory. It is also known as sequential representation. Sequential representation of binary trees is done using one-dimensional (1-D) arrays. This type of representation is static and hence inefficient, as the size must be known in advance and thus requires a lot of memory space. The following rules are used to decide the location of each node in the memory:

(a) The root node of the tree is stored in the first location.

(b) If the parent node is present at location k, then the left child is stored at location 2k, and the right child is stored at location (2k + 1).

(c) The maximum size of the array is given as $(2^h - 1)$, where h is the height of the tree.

For example: A binary tree is given as follows. Give its array representation in the memory.

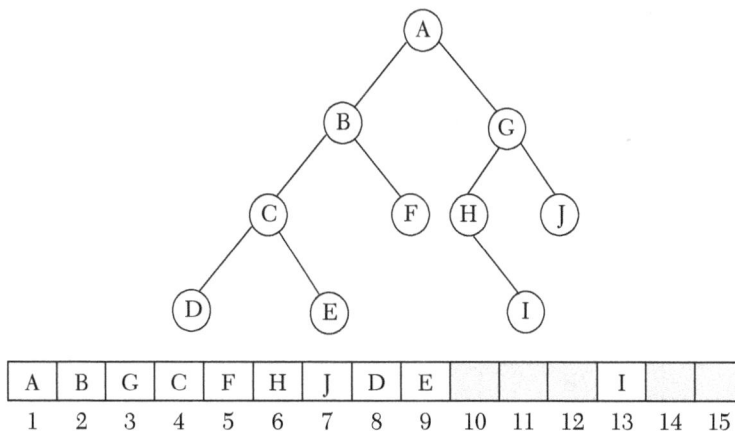

Figure 8.12. Binary tree and its array representation.

Linked Representation of Binary Trees

A binary tree can also be represented using a linked list in a computer's memory. This type of representation is dynamic, as memory is dynamically allocated, that is, when it is needed, and thus it is efficient and avoids wastage of memory space. In linked representation, every node has three parts:

1. The first part is called the left child, which contains the address of the left subtree.

2. The second part is called the data part, which contains the information of the node.

3. The third part is called the right child, which contains the address of the right subtree.

The class of the node is declared as follows:

```
private class Node {
        int data;
        Node left;
        Node right;
}
```

The representation of a node is given in Figure 8.2. When there are no children of a node, the corresponding fields are NULL.

For example: A binary tree is given as follows. Give its linked representation in the memory.

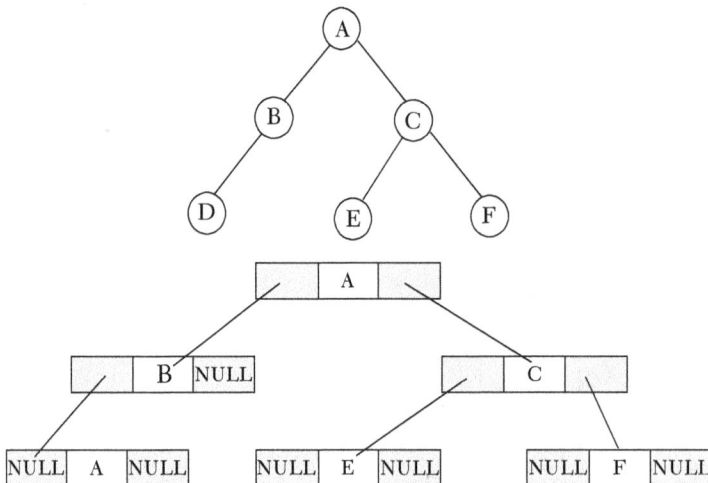

Figure 8.13. Binary tree and its linked representation.

8.4   Binary Search Tree

A binary search tree (BST) is a variant of a binary tree. *The special property of a binary search tree is that all the nodes in the left subtree have a value less than that of the root node. Similarly, all the nodes in the right subtree have a value more than that of the root node.* Hence, the binary search tree is also known as an ordered binary tree, because all the nodes in a binary search tree are ordered. Also, the left and the right subtrees are also binary search trees, and thus the same property is applicable on every subtree in the binary search tree. Figure 8.14 represents a binary search tree in which all the keys are ordered.

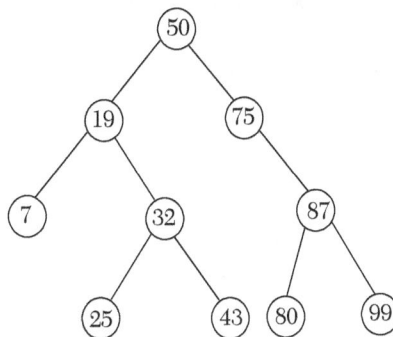

Figure 8.14. Binary search tree.

In the previous figure, the root node is 50. The left subtree of the root node consists of the nodes 19, 7, 32, 25, and 43. We can see that all these nodes have smaller values than the root node, and hence it constitutes the left subtree. Similarly, the right subtree of the root node consists of the nodes 75, 87, 80, and 99. Here also we can see that all these nodes have higher values than the root node, and hence it constitutes the right subtree. Also, each of the subtrees is ordered. Thus, it becomes easier to search for an element in the tree and, as a result, time is also reduced by a great margin. *Binary search trees are very efficient regarding searching for an element.* These trees are already sorted in nature. Thus, these trees have a low time complexity. Various operations which can be performed on binary search trees will be discussed in the upcoming section.

8.4.1   Operations on Binary Search Trees

In this section, we will discuss different operations that are performed on binary search trees, which include:

▪ Searching for a node/key in the binary search tree

▪ Inserting a node/key in the binary search tree

- Deleting a node/key from the binary search tree

- Deleting the entire binary search tree

- Finding the mirror image of the binary search tree

- Finding the smallest node in the binary search tree

- Finding the largest node in the binary search tree

- Determining the height of the binary search tree

Now, let us discuss all of these operations in detail.

1. Searching for a node/key in the binary search tree: The searching operation is one of the most common operations performed in the binary search tree. *This operation is performed to find whether a given key exists in the tree or not.* The searching operation starts at the root node. First, it will check whether the tree is empty or not. If the tree is empty, then the node/key for which we are searching is not present in the tree, and the algorithm terminates there by displaying the appropriate message. If the tree is not empty and the nodes are present in it, then the search function checks the node/value to be searched and compares it with the key value of the current node. If the node/key to be searched is less than the key value of the current node, then in that case, we will recursively call the left child node. On the other hand, if the node/key to be searched is greater than the key value of the current node, then we will recursively call the right child node. Now, let us look at the algorithm for searching for a key in a binary search tree.

Algorithm for searching for a node/key in a binary search tree

```
SEARCH(ROOT, VALUE)
Step 1: START
Step 2: IF(ROOT == NULL)
            Return NULL
            Print "Empty Tree"
        ELSE IF(ROOT.INFO == VALUE)
            Return ROOT
        ELSE IF(ROOT.INFO > VALUE)
            SEARCH(ROOT.LCHILD, VALUE)
        ELSE IF(ROOT.INFO < VALUE)
            SEARCH(ROOT.RCHILD, VALUE)
        ELSE
            Print "Value not found"
        [End of IF]
    [End of IF]
[End of IF]
```

[End of IF]

Step 3: END

In the previous algorithm, first we check whether the tree is empty or not. If the tree is empty, then we return NULL. If the tree is not empty, then we check whether the value stored at the current node (ROOT) is equal to the node/key we want to search or not. If the value of the ROOT node is equal to the key value to be searched, then we return the current node of the tree, that is, the ROOT node. Otherwise, if the key value to be searched is less than the value stored at the current node, we recursively call the left subtree. If the key value to be searched is greater than the value stored at the current node, then we recursively call the right subtree. Finally, if the value is not found, then an appropriate message is printed on the screen.

For example: We have been given a binary search tree. Now, search for the node with the value 20 in the binary search tree.

Initially, the binary search tree is given as:

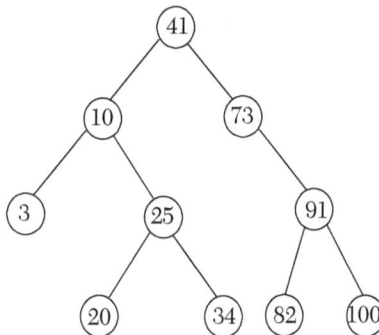

Figure 8.15(a)

Step 1: First, the root node, that is, 41, is checked.

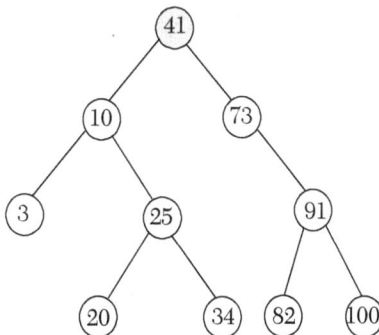

Figure 8.15(b)

Step 2: Second, as the value stored at the root node is not equal to the value we are searching for, but we know that 20 < 41, thus we will traverse the left subtree.

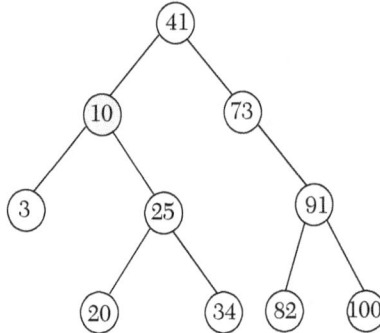

Figure 8.15(c)

Step 3: We know that 10 is not the value to be searched, but 20 > 10; thus, we will now traverse the right subtree with respect to 10.

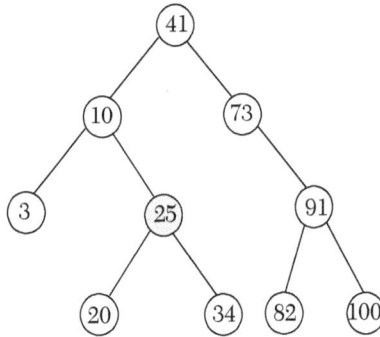

Figure 8.15(d)

Step 4: Again 25 is not the value to be searched, but 20 < 25; thus, we will now traverse the left subtree with respect to 25.

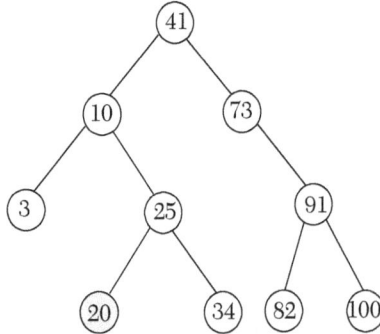

Figure 8.15. Searching a node with value 20 in the binary search tree.

Finally, a node having value 20 is successfully searched for in the binary search tree.

2. Inserting a node/key in the binary search tree: The insertion operation is performed to insert a new node with the given value in the binary search tree. The new node is inserted at the correct position following the binary search tree constraint. It should not violate the property of the binary search tree. The insertion operation also starts at the root node. First, it will check whether the tree is empty or not. If the tree is empty, then we will allocate the memory for the new node. If the tree is not empty, then we will compare the key value to be inserted with the value stored in the current node. If the node/key to be inserted is less than the key value of the current node, then the new node is inserted in the left subtree. On the other hand, if the node/key to be inserted is greater than the key value of the current node, then the new node is inserted in the right subtree. Now, let us discuss the algorithm for inserting a node in a binary search tree.

Algorithm for inserting a node/key in a binary search tree

```
INSERT(ROOT, VALUE)
Step 1: START
Step 2: IF(ROOT == NULL)
            Allocate memory for ROOT node
            Set ROOT.INFO = VALUE
            Set ROOT.LCHILD = ROOT.RCHILD = NULL
        [End of IF]
Step 3: IF(ROOT.INFO.VALUE)
            INSERT(ROOT.LCHILD, VALUE)
        ELSE
            INSERT(ROOT.RCHILD, VALUE)
        [End of IF]
Step 4: END
```

In the previous algorithm, first we check whether the tree is empty or not. If the tree is empty, then we will allocate memory for the ROOT node. In Step 3, we are checking whether the key value to be inserted is less than the value stored at the current node; if so, we will simply insert the new node in the left subtree. Otherwise, the new child node is inserted in the right subtree.

For example: We have been given a binary search tree. Now, insert a new node with the value 7 in the binary search tree.

Initially, the binary search tree is given as:

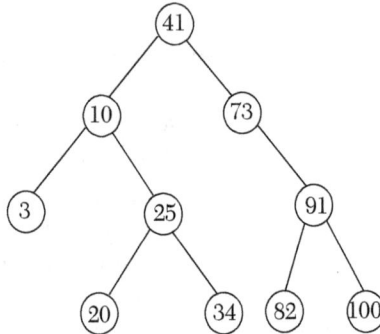

Figure 8.16(a)

Step 1: First, we check whether the tree is empty or not. So, we will check the root node. As the root node is not empty, we will begin the insertion process.

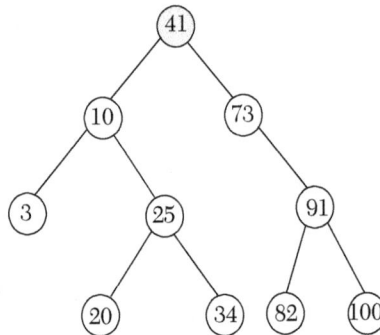

Figure 8.16(b)

Step 2: Second, we know that 7 < 41; thus, we will traverse the left subtree to insert the new node.

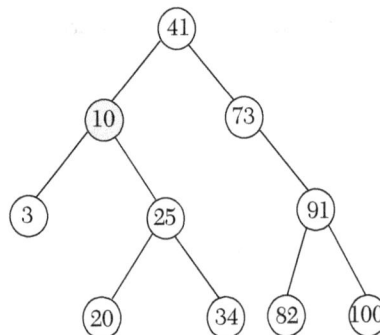

Figure 8.16(c)

Step 3: Third, we know that 7 < 10; thus, we will again traverse the left subtree to insert the new node.

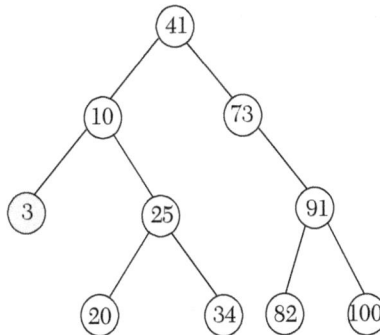

Figure 8.16(d)

Step 3: Now, we know that 7 > 3, thus the new node with value 7 is inserted as the right child of the parent node 3.

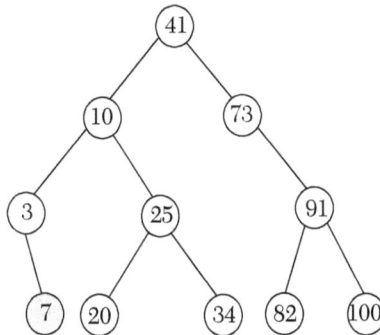

Figure 8.16: Inserting a new node with value 7 in the binary search tree.

Finally, the new node with the value 7 is inserted as a right child in the binary search tree.

3. Deleting a node/key from a binary search tree: Deleting a node/key from a binary search tree is the most crucial process. We should be careful when performing the deletion operation; while deleting the nodes, we must be sure that the property of the binary search tree is not violated so that we don't lose necessary nodes during this process. The deletion operation is divided into three cases as follows:

Case 1: Deleting a node having no children

This is the simplest case of deletion, as we can directly remove or delete a node which has no children. Look at the binary search tree given in Figure 8.17 and see how deletion is done in this case.

For example: We have been given a binary search tree. Now, delete a node with the value 61 from the binary search tree.

Initially the binary search tree is given as:

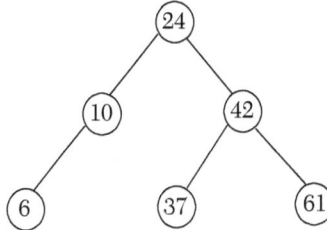

Figure 8.17(a)

Step 1: First, we will check whether the tree is empty or not by checking the root node.

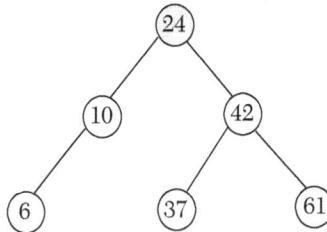

Figure 8.17(b)

Step 2: Second, as the root node is present, we will compare the value to be deleted with the value stored at the current node. As 61 > 24, we will recursively traverse the right subtree.

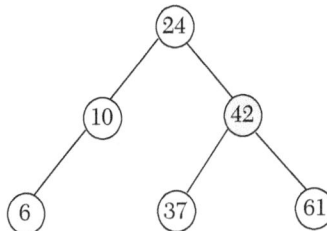

Figure 8.17(c)

Step 3: Again, we will compare the value to be deleted with the value stored at the current node. As 61 > 42, we will recursively traverse the right subtree.

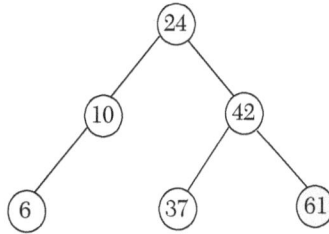

Figure 8.17(d)

Step 4: Finally, a node having value 61 is deleted from the binary search tree.

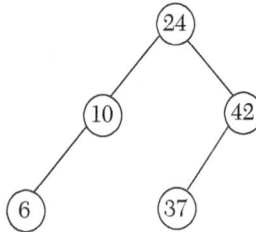

Figure 8.17. Deleting the node with value 61 from the binary search tree.

Case 2: Deleting a node having one child

In this case of deletion, the node which is to be deleted, the parent node, is simply replaced by its child node. Look at the binary search tree given in Figure 8.18 and see how deletion is done in this case.

For example: We have been given a binary search tree. Now, delete a node with the value 10 from the binary search tree.

Initially the binary search tree is given as:

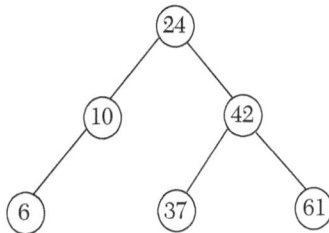

Figure 8.18(a)

Step 1: First, we will check whether the tree is empty or not by checking the root node.

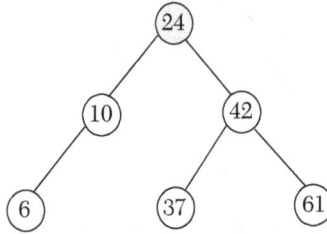

Figure 8.18(b)

Step 2: Second, as the root node is present, we will compare the value to be deleted with the value stored at the current node. As 10 < 24, we will recursively traverse the left subtree.

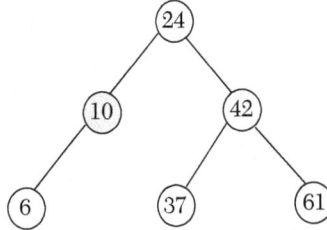

Figure 8.18(c)

Step 3: Now, as the node to be deleted is found and has one child, the node to be deleted is now replaced by its child node, and the actual node is deleted.

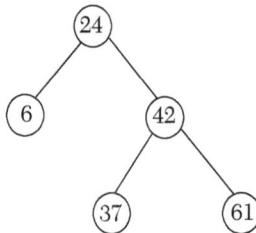

Figure 8.18. Deleting the node with value 10 from the binary search tree.

Case 3: Deleting a node having two children

In this case, the node which is to be deleted is simply replaced by its in-order predecessor, that is, the largest value in the left subtree, or by its in-order successor, that is, the smallest value in the right subtree. Also, the in-order predecessor or in-order successor can be deleted using any of the two cases. Now, look at the binary search tree shown in Figure 8.19 and see how the deletion will take place in this case.

Let us discuss the algorithm for deleting a node from a binary search tree.

For example: We have been given a binary search tree. Now, delete a node with the value 42 from the binary search tree.

Initially, the binary search tree is given as:

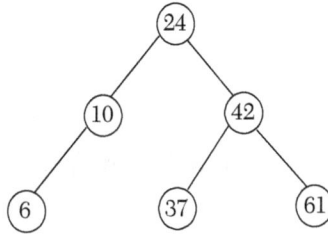

Figure 8.19(a)

Step 1: First, we will check whether the tree is empty or not by checking the root node.

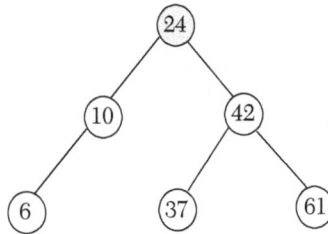

Figure 8.19(b)

Step 2: Second, as the root node is present, we will compare the value to be deleted with the value stored at the current node. As 42 > 24, we will recursively traverse the right subtree.

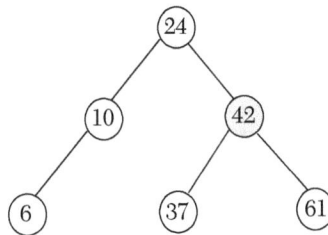

Figure 8.19(c)

Step 3: As the node to be deleted is found and has two children, now we will find the in-order predecessor of the current node (42) and replace the current node with its in-order predecessor so that the actual node 42 is deleted.

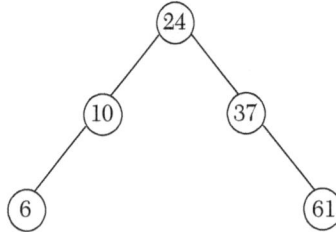

Figure 8.19. Deleting the node with value 42 from the binary search tree.

Algorithm for deleting a node/key from a binary search tree

```
DELETE_NODE(ROOT, VALUE)
Step 1: START
Step 2: IF(ROOT == NULL)
        Print "Error"
        [End of IF]
Step 3: IF(ROOT.INFO > VALUE)
            DELETE_NODE(ROOT.LCHILD, VALUE)
        ELSE IF(ROOT.INFO < VALUE)
            DELETE_NODE(ROOT.RCHILD, VALUE)
        ELSE IF(ROOT.LCHILD = NULL & ROOT.RCHILD = NULL)
            ROOT = NULL
        ELSE
           IF(ROOT.LCHILD & ROOT.RCHILD)
           TEMP = FIND_LARGEST(ROOT.LCHILD)
        OR
           TEMP = FIND_SMALLEST(ROOT.RCHILD)
           Set ROOT.INFO = TEMP.INFO
        ELSE
           IF(ROOT.LCHILD != NULL)
           Set TEMP = ROOT.LCHILD
           Set ROOT.INFO = TEMP.INFO
        ELSE
           Set TEMP = ROOT.RCHILD
           Set ROOT.INFO = TEMP.INFO
        [End of IF]
     [End of IF]
        [End of IF]
Step 4: END
```

In the previous algorithm, first we check whether the tree is empty or not. If the tree is empty, then the node to be deleted is not present. Otherwise, if the tree is not empty, we will check whether the node/value to be deleted is less than the value stored at the current node. If the value to be deleted is less, then we will recursively call the left subtree. If the value to be deleted is greater than the value stored at the current node, then we will recursively call the right subtree. Now, if the node to be deleted has no children, then the node is simply freed. If the node to be deleted has two children, that is, both a left and right child, then we will find the in-order predecessor by calling (`TEMP = FIND_LARGEST(ROOT.LCHILD)` or in-order successor by calling (`TEMP = FIND_SMALLEST(ROOT.RCHILD)` and replace the value stored at the current node with that of the in-order predecessor or in-order successor. Then, we will simply delete the initial node of either the in-order predecessor or in-order successor. Finally, if the node to be deleted has only one child, the value stored at the current node is replaced by its child node and the child node is deleted.

4. Deleting the entire binary search tree: It is very easy to delete the entire binary search tree. First, we will delete all the nodes present in the left subtrees followed by the nodes present in the right subtree. Finally, the root node is deleted, and the entire tree is deleted.

Algorithm for deleting an entire binary search tree

```
DELETE_BST(ROOT)
Step 1: START
Step 2: IF(ROOT != NULL)
            DELETE_BST(ROOT.LCHILD)
            DELETE_BST(ROOT.RCHILD)
            ROOT = NULL
        [End of IF]
Step 3: END
```

5. Finding the mirror image of a binary search tree: This is an exciting operation to perform in a binary search tree. The mirror image of the binary search tree means interchanging the right subtree with the left subtree at each and every node of the tree.

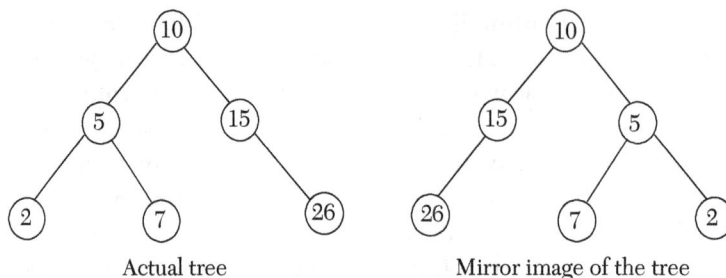

Actual tree   Mirror image of the tree

Figure 8.20. Binary search tree and its mirror image.

Algorithm for finding the mirror image of a binary search tree

```
MIRROR_IMAGE (ROOT)
Step 1: START
Step 2: IF(ROOT != NULL)
            MIRROR_IMAGE(ROOT.LCHILD)
            MIRROR_IMAGR(ROOT.RCHILD)
            Set TEMP = ROOT.LEFT
            ROOT.LEFT = ROOT.RIGHT
            Set ROOT.RIGHT = TEMP
        [End of IF]
Step 3: END
```

6. Finding the smallest node in the binary search tree: We know that it is the basic property of the binary search tree that the smallest value always occurs in the extreme left of the left subtree. If there is no left subtree, then the value of the root node will be the smallest. Hence, to find the smallest value in the binary search tree, we will simply find the value of the node present at the extreme left of the left subtree.

Algorithm for finding the smallest node in abinary search tree

```
SMALLEST_VALUE (ROOT)
Step 1: START
Step 2: IF(ROOT = NULL OR ROOT.LCHILD = NULL)
           Return ROOT
        ELSE
           Return SMALLEST_VALUE(ROOT.LCHILD)
        [End of IF]
Step 3: END
```

7. Finding the largest node in a binary search tree: We know that it is the basic property of the binary search tree that the largest value always occurs in the extreme right of the right subtree. If there is no right subtree, the value of the root node will be the largest. Hence, to find the largest value in a binary search tree, we will simply find the value of the node present at the extreme right of the right subtree.

Algorithm for finding the largest node in a binary search tree

```
LARGEST_VALUE(ROOT)
Step 1: START
Step 2: IF(ROOT = NULL OR ROOT.RCHILD = NULL)
           Return ROOT
        ELSE
           Return LARGEST_VALUE(ROOT.RCHILD)
        [End of IF]
Step 3: END
```

8. Determining the height of a binary search tree: The height of a binary search tree can easily be determined. We will first calculate the heights of the left subtree and the right subtree. Whichever height is greater, 1 is added to that height; that is, if the height of the left subtree is greater, then 1 is added to the height of the left subtree. Similarly, if the height of the right subtree is greater, then 1 is added to the height of the right subtree.

Algorithm for determining the height of a binary search tree

```
CALCULATE_HEIGHT(ROOT)
Step 1: START
Step 2: IF ROOT = NULL
           Print "Can't find height of the tree."
        ELSE
           Set LHEIGHT = CALCULATE_HEIGHT(ROOT.LCHILD)
           Set RHEIGHT = CALCULATE_HEIGHT(ROOT.RCHILD)
        IF(LHEIGHT < RHEIGHT)
        Return (RHEIGHT) + 1
        ELSE
           Return (LHEIGHT) + 1
         [End of IF]
         [End of IF]
Step 3: END
```

8.4.2 Binary Tree Traversal Methods

Traversing is the process of visiting each node in the tree exactly once in a particular order. We all know that a tree is a non-linear data structure, and therefore a tree can be traversed in various ways. There are three types of traversals, which are:

 ▪ Pre-Order Traversal

▪ In-Order Traversal

▪ Post-Order Traversal

Now, we will discuss all of these traversals in detail.

Pre-Order Traversal

In pre-order traversal, the following operations are performed recursively at each node:

1. Visit the root node.

2. Traverse the left subtree.

3. Traverse the right subtree.

The word "pre" in pre-order determines that the root node is accessed before accessing any other node in the tree. Hence, it is also known as a DLR traversal, that is, Data Left Right. Therefore, in a DLR traversal, the root node is accessed first followed by the left subtree and right subtree. Now, let us see an example for pre-order traversal.

For example: Find the pre-order traversal of the given binary tree of the word EDUCATION.

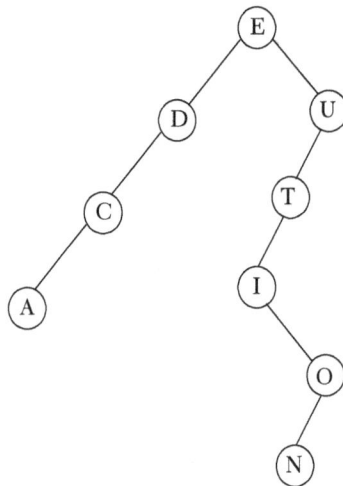

The pre-order traversal of the previous binary tree is:

| E D C A U T I O N |

Now, let us look at the function for pre-order traversal.

Function for pre-order traversal

```
public void preorder() {
    preorder(root);
    System.out.println();
}
private void preorder(Node node) {
    if (node == null)
        return;
    System.out.print(node.data + " ");
    preorder(node.left);
    preorder(node.right);
}
```

In-Order Traversal

In in-order traversal, the following operations are performed recursively at each node:

1. Traverse the left subtree.

2. Visit the root node.

3. Traverse the right subtree.

The word "in" in "in-order" determines that the root node is accessed in between the left and the right subtrees. Hence, it is also known as an LDR traversal, that is, Left Data Right. Therefore, in an LDR traversal, the left subtree is traversed first followed by the root node and the right subtree. Now, let us see an example for an in-order traversal.

For example: Find the in-order traversal of the given binary tree of the word EDUCATION.

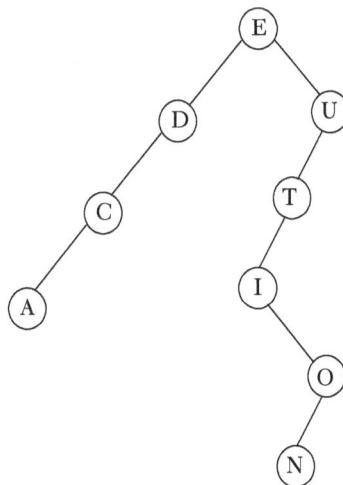

The in-order traversal of the previous binary tree is:

A C D E I N O T U

Now, let us look at the function for an in-order traversal.

Function for an in-order traversal

```
public void inorder() {
    inorder(root);
    System.out.println();
}
private void inorder(Node node) {
    if (node == null)
        return;
    inorder(node.left);
    System.out.print(node.data + " ");
    inorder(node.right);
}
```

Post-Order Traversal

In a post-order traversal, the following operations are performed recursively at each node:

1. Traverse the left subtree.

2. Traverse the right subtree.

3. Visit the root node.

The word "post" in post-order determines that the root node will be accessed last after the left and the right subtrees. Hence, it is also known as an LRD traversal, that is, Left Right Data. Therefore, in an LRD traversal, the left subtree is traversed first followed by the right subtree and the root node. Now, let us see an example for a post-order traversal.

For example: Find the post-order traversal of the given binary tree of the word EDUCATION.

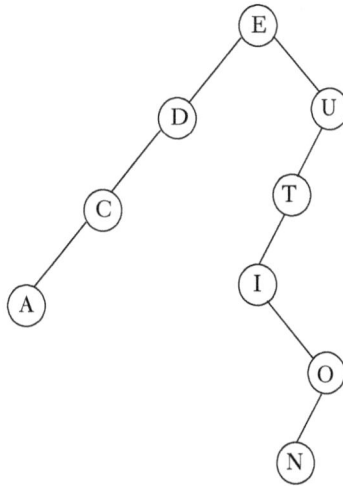

The post-order traversal of the previous binary tree is:

A C D N O I T U E

Now, let us look at the function of the post-order traversal.

Function for post-order traversal

```
public void postorder() {
    postorder(root);
    System.out.println();
}
private void postorder(Node node) {
    if (node == null)
        return;
    postorder(node.left);
    postorder(node.right);
    System.out.print(node.data + " ");
}
```

//Write a program to create a binary search tree and perform different operations on it.

```
public class BinarySearchTree {
    private class Node {
        int data;
        Node left;
        Node right;
    }
    private Node root;
```

```java
public void Search(int item) {
    boolean ans = Search(root, item);
    if(ans) {
        System.out.println(item + " found in tree");
    }else {
System.out.println(item + " not found in tree");
    }
}
private boolean Search(Node node, int item) {
    if (node == null) {
        return false;
    }
    if (item < node.data) {
        return Search(node.left, item);
    } else if (item > node.data) {
        return Search(node.right, item);
    } else {
        return true;
    }
}
public void insert(int item) {
    if (root == null) {
        Node nn = new Node();
        nn.data = item;
        root = nn;
    } else
        insert(root, item);
    System.out.println("Insertion successful!!!");
    System.out.print("Tree after insertion is: ");
    display();
    System.out.println();
}
private void insert(Node node, int item) {
    if (node == null) {
        return;
    }
    if (item <= node.data) {
        if (node.left == null) {
            Node nn = new Node();
            nn.data = item;
            node.left = nn;
            return;
        }
```

```
            insert(node.left, item);
        } else {
            if (node.right == null) {
                Node nn = new Node();
                nn.data = item;
                node.right = nn;
                return;
            }
            insert(node.right, item);
        }
    }
    public void delete(int item) {
        if (root.data == item) {
            if (root.left == null) {
                root = root.right;
            } else if (root.right == null) {
                root = root.left;
            } else {
                delete(root, null, item);
            }
        } else
            delete(root, null, item);
        System.out.println("Deletion successful!!!");
        System.out.print("Tree after deletion is: ");
        display();
        System.out.println();
    }
    private void delete(Node node, Node parent, int item) {
        if (node == null) {
            return;
        }
        if (item > node.data) {
            delete(node.right, node, item);
        } else if (item < node.data) {
            delete(node.left, node, item);
        } else {
            if (node.left == null && node.right == null) { // case1
                if (item <= parent.data)
                    parent.left = null;
                else
                    parent.right = null;
            } else if (node.left != null && node.right == null)
                    { // case2
```

```java
                if (item <= parent.data)
                    parent.left = node.left;
                else
                    parent.right = node.left;
        } else if (node.left == null && node.right != null)
                { // case3
                if (item <= parent.data)
                    parent.left = node.right;
                else
                    parent.right = node.right;
        } else { // case4
                int max = max(node.left);
                delete(node.left, node, max);
                node.data = max;
            }
        }
    }
    public void min() {
        System.out.println("Min element of tree is: " + min(root));
    }
    private int min(Node node) {
        if (node.left == null) {
            return node.data;
        }
        return min(node.left);
    }
    public void max() {
        System.out.println("Max element of tree is: " + max(root));
    }
    private int max(Node node) {
        if (node.right == null) {
            return node.data;
        }
            return max(node.right);
    }
    public void height() {
        System.out.println("Height of tree is: " + height(root));
    }
    private int height(Node node) {
        if (node == null) {
            return -1;
        }
        int ht = 0;
        int lh = height(node.left);
```

```
        int rh = height(node.right);
        if (lh > rh) {
            ht += lh;
        } else {
            ht += rh;
        }
        return ht + 1;
}
public void preorder() {
    preorder(root);
    System.out.println();
}
private void preorder(Node node) {
    if (node == null)
        return;
    System.out.print(node.data + " ");
    preorder(node.left);
    preorder(node.right);
}
public void postorder() {
    postorder(root);
    System.out.println();
}
private void postorder(Node node) {
    if (node == null)
        return;
    postorder(node.left);
    postorder(node.right);
    System.out.print(node.data + " ");
}
    public void inorder() {
    inorder(root);
    System.out.println();
}
private void inorder(Node node) {
    if (node == null)
        return;
    inorder(node.left);
    System.out.print(node.data + " ");
    inorder(node.right);
}
public void display() {
    inorder(root); //displays the inorder traversal of the tree
```

```
            System.out.println();
        }
}
//Client Class
public class BinarySearchTreeClient {
    public static void main(String[] args) {
            BinarySearchTree bst = new BinarySearchTree();
            bst.insert(60);
            bst.insert(20);
            bst.insert(90);
            bst.insert(40);
            bst.insert(50);
            bst.insert(80);
            bst.delete(80);
            bst.delete(90);
            bst.height();
            bst.max();
            bst.min();
            bst.Search(30);
            System.out.println();
            System.out.println("Inorder Traversal");
            bst.inorder();
            System.out.println("Preorder Traversal");
            bst.preorder();
            System.out.println("Postorder Traversal");
            bst.postorder();
        }
}
```

The output of the program is shown as:

8.4.3 Creating a Binary Tree Using Traversal Methods

A binary tree can be constructed if we are given at least two of the traversal results, provided that one traversal should always be an in-order traversal and the second is either a pre-order traversal or a post-order traversal. An in-order traversal determines the left and right child nodes of the binary tree. A Pre-order or post-order traversal determines the root node of the binary tree. Hence, there are two different ways of creating a binary tree, which are:

1. In-order and pre-order traversal

2. In-order and post-order traversal

Now, we have pre-order and in-order traversal sequences. Then, the following steps are followed to construct a binary tree:

Step 1: The pre-order traversing sequence is used to determine the root node of the binary tree. The first node in the pre-order sequence will be the root node.

Step 2: The in-order traversing sequence is used to determine the left and the right subtrees of the binary tree. Keys toward the left side of the root node in the in-order sequence form the left subtree. Similarly, keys toward the right side of the root node in the in-order sequence form the right subtree.

Step 3: Now, each element from the pre-order traversing sequence is recursively selected, and the left and the right subtrees are created from the in-order traversing sequence.

For example: Create a binary tree from the given traversing sequences.

In-order: A C D E I N O T U

Pre-order: E DCAU T I O N

Now, we will construct the binary tree.

1. The first node in the pre-order sequence is the root node of the tree. Hence, E is the root node of the binary tree.

(E)
Root node

2. Now, we can easily determine the left and right subtrees from the in-order sequence. Keys toward the left side of the root node, that is, A, C, and D, form the left subtree. Similarly, elements on the right side of the root node, that is, I, N, O, T, and U, form the right subtree.

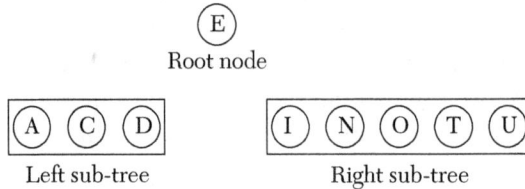

(E)
Root node

(A) (C) (D)   (I) (N) (O) (T) (U)

Left sub-tree   Right sub-tree

3. Now, the left child of the root node will be the first node in the pre-order traversing sequence after the root node E. Thus, D is the left child of the root node E.

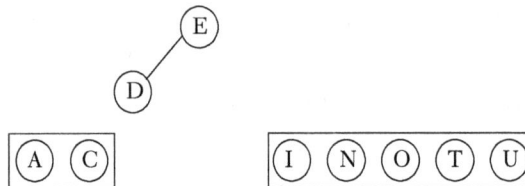

(E)
(D)

(A) (C)   (I) (N) (O) (T) (U)

4. Similarly, the right child of the root node will be the first node in the pre-order traversing sequence after the nodes of the left subtree. Thus, U is the right child of the root node E.

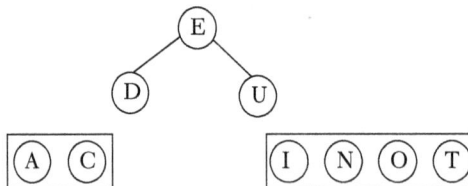

(E)
(D)   (U)

(A) (C)   (I) (N) (O) (T)

5. In the in-order sequence, A and C are on the left side of D. So, A and C will form the left subtree of D.

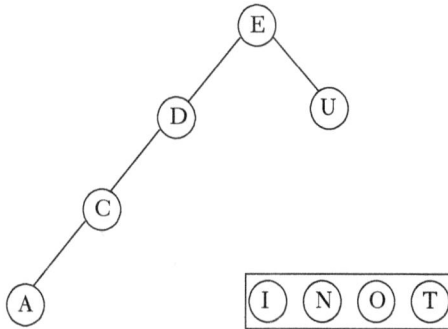

6. Now, the next elements in the pre-order sequence are T and I. Also, in the in-order sequence, T and I are on the left side of U. So, T and I will form the left subtree of U.

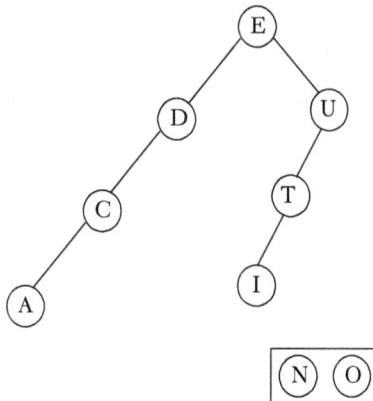

7. The next element in the pre-order sequence is O. In the in-order sequence, O is on the right side of I. So, O will form the right subtree of I. The last element in the pre-order sequence is N. N is on the left side of O in the in-order sequence. Thus, N will form the left subtree of O.

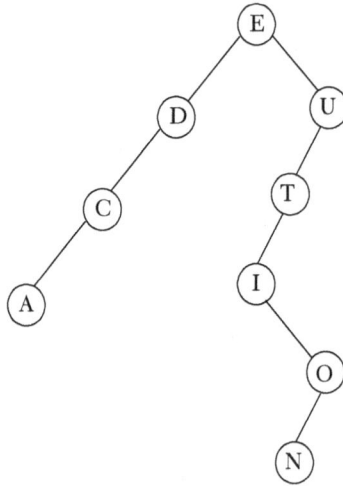

Finally, the binary tree is created from the given traversing sequences.

Frequently Asked Questions

1. Create a binary tree from the given traversing sequences.

In-order – d b e a f c g
Pre-order – a b d e c f g
      Ans:

Step 1: a is the root node of the binary tree.

Step 2: d, b, and e are on the left side of the a node in the in-order sequence. Hence, d, b, and e are the left subtrees of root a. Also, d is the left subtree of b, and e is the right subtree of b.

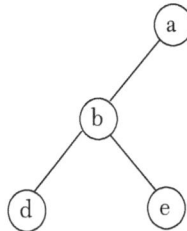

Step 3: f, c, and g are on the right side of root a in the in-order sequence. Hence, f, c, and g are the right subtrees of root a. Also, f is the left subtree of c, and g is the right subtree of c.

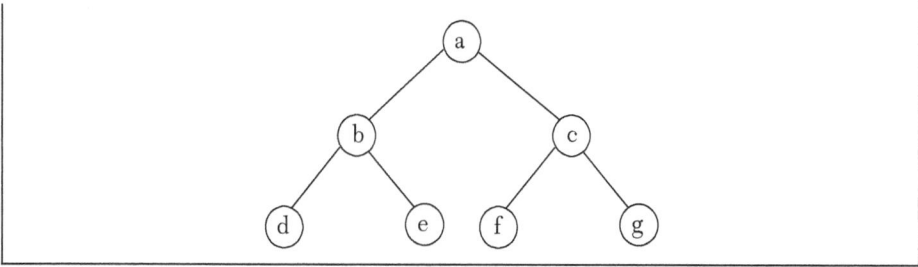

8.5   AVL Trees

The AVL tree was invented by Adelson-Velski and Landis in 1962 and was named in honor of its inventors. The AVL tree was the first balanced binary search tree. It is a self-balancing binary search tree. The AVL tree is also known as a height-balanced tree because of its property that the heights of the two subtrees of a node can differ at most by one. AVL trees are very efficient in performing searching, insertion, and deletion operations, as they take O(log n) time to perform all these operations.

8.5.1   Need of Height-Balanced Trees

AVL trees are very similar to binary search trees but with a small difference. AVL trees have a special variable known as a balance factor associated with them. Every node in the AVL tree has a balance factor associated with it. The balance factor is determined by subtracting the height of the right subtree from the height of the left subtree. Thus, *a node with a balance factor of -1, 0, or 1 is said to be a height-balanced tree.* The primary need for the height-balanced tree is that the process of searching becomes very fast. This balancing condition also ensures that the depth of the tree is O(log n). The balance factor is calculated as follows:

Balance Factor = Height(Left subtree) – Height(Right subtree)

- If the balance factor of the tree is -1, then it means that the height of the right subtree of that node is one more than the height of the left subtree of that node.

- If the balance factor of the tree is 0, then it means that the height of the left and the right subtrees of a node are equal.

- If the balance factor of the tree is 1, then it means that the height of the left subtree of that node is one more than the height of its right subtree.

Thus, the overall benefit of the height-balanced tree is to assist in fast searching.

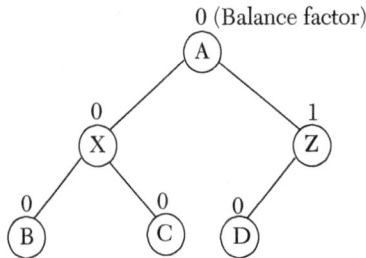

Figure 8.21. Balanced AVL tree.

8.5.2 Operations on an AVL Tree

In this section, we will discuss various operations which are performed on AVL trees. These are:

▪ Searching a node in an AVL Tree

▪ Inserting a new node in an AVL Tree

Now, let us discuss both of them in detail.

1. Searching a node in an AVL Tree

The process of searching a node in an AVL tree is the same as for a binary search tree.

2. Inserting a new node in an AVL Tree

The process of inserting a new node in an AVL tree is quite similar to that of binary search trees. The new node is always inserted as a terminal/leaf node in the AVL tree. But the insertion of a new node can disturb the balance of the AVL tree, as the balance factor may be disturbed. Thus, for the tree to remain balanced, the insertion process is followed by a rotation process. The rotation process is usually done to restore the balance factor of the tree. If the balance factor of each node is -1, 0, or 1 after the insertion process, then rotation is not required, as the tree is already balanced; otherwise, rotation is required. Now, let us look at the given example and see how insertion is done without rotations.

For example: In the given AVL tree, insert a new node with value 60 in the tree.

Initially, the AVL tree is given as:

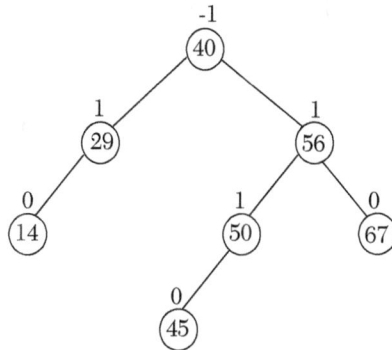

Figure 8.22. AVL tree before insertion.

Now, we will insert 60 into the AVL tree.

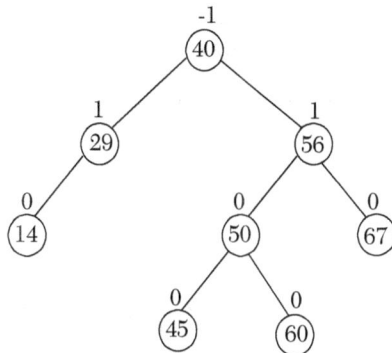

Figure 8.23. AVL tree after inserting 60.

Hence, after insertion, there are no nodes in the tree which are unbalanced. Thus, there is no need to apply rotation here. However, now we will discuss how the rotation process is performed in AVL trees.

AVL Rotations

Rotation is done when the balance factor of the node becomes disturbed after inserting a new node. We know that the new node which is inserted will always have a balance factor of 0, as it will be a leaf node. Hence, the nodes whose balance factors will be disturbed are the ones which lie in the path of the root node to the newly inserted node. So, we will perform the rotation process only on those nodes whose balance factors will be disturbed. In the rotation process, our first work is to find the critical node in the AVL tree. The critical node is the nearest ancestor node from the newly inserted node to the root node which does not has a balance factor of -1, 0, or 1. First, let us understand the concept of the critical node with the help of an example.

For example: Find the critical node in the given AVL tree.

Initially, the AVL tree is given as follows:

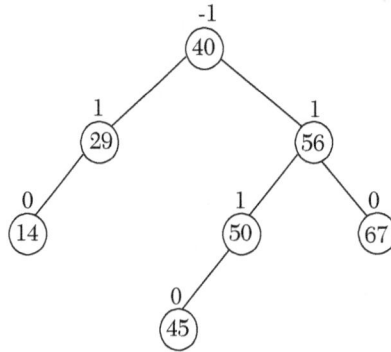

Figure 8.24. AVL tree.

Now, we will insert a new node with value 42 in the tree.

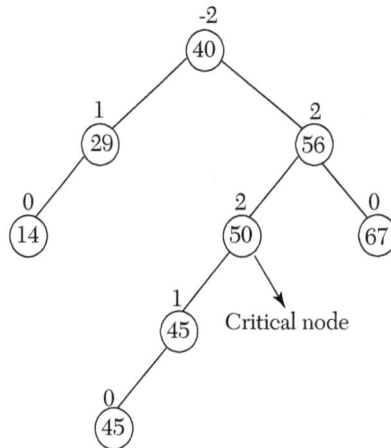

Figure 8.25. AVL tree.

After inserting 42 in the AVL tree, we can see that there are three nodes in the tree which have balance factors equal to -2, 2, and 2. Now, the critical node is the one which is the nearest to the newly inserted node with a disturbed balance factor. We can see that 50 is the nearest node to 42, and 50 has a balance factor of 2. Thus, 50 is the critical node in this AVL tree.

However, to restore the balance factor of the previous AVL tree, rotations are performed. There are four types of rotations which are:

1. **Left-Left Rotation (LL Rotation):** New node is inserted in the left subtree of the left subtree of the critical node.

2. **Right-Right Rotation (RR Rotation):** New node is inserted in the right subtree of the right subtree of the critical node.

3. **Right-Left Rotation (RL Rotation):** New node is inserted in the left subtree of the right subtree of the critical node.

4. **Left-Right Rotation (LR Rotation):** New node is inserted in the right subtree of the left subtree of critical node.

Now, let us discuss all of these rotations in detail.

LL Rotation

LL rotation is also known as Left-Left rotation, as the new node is inserted in the left subtree of the left subtree of the critical node. It is a single rotation. Let us take an example and perform an LL rotation in it.

For example:

Initially, the AVL tree is given as:

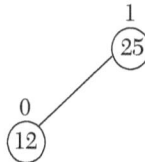

Figure 8.26(a)

Insert new node 5 in the AVL tree.

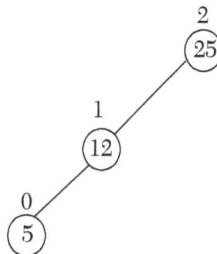

Figure 8.26(b)

After inserting 5 in the AVL tree, the balance factor of 25 is disturbed. Thus, 25 is the critical node. Hence, we will apply LL rotation to restore the balance factor of the tree. After rotation node 12 becomes the root node, node 5 and node 25 become the left and the right child of the tree respectively.

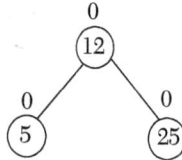

Figure 8.26. Showing an LL rotation in an AVL tree.

Therefore, the LL rotation is performed, and the balance factor of each node is also restored.

RR Rotation

RR rotation is also known as Right-Right rotation, as the new node is inserted in the right subtree of the right subtree of the critical node. It is also a single rotation. Let us take an example and perform an RR rotation in it.

For example:

Initially the AVL tree is given as follows:

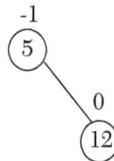

Figure 8.27(a)

Insert new node 25 in the AVL tree.

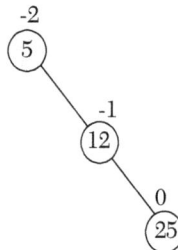

Figure 8.27(b)

After inserting 25 in the AVL tree, the balance factor of 5 is disturbed. Thus, 5 is the critical node. Hence, here we will apply an RR rotation to restore the balance factor of the tree. After rotation node 12 becomes the root node, node 5 and node 25 become the left and the right child of the tree respectively.

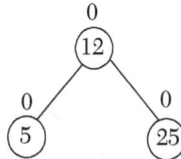

Figure 8.27. Showing an RR rotation in an AVL tree.

Therefore, the RR rotation is performed, and the balance factor of each node is also restored.

RL Rotation

RL rotation is also known as Right-Left rotation, as the new node is inserted in the left subtree of the right subtree of the critical node. It is a double rotation. Let us take an example and perform an RL rotation in it.

For example:

Initially, the AVL tree is given as follows:

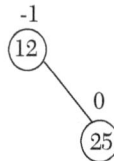

Figure 8.28(a)

Insert new node 15 in the AVL tree.

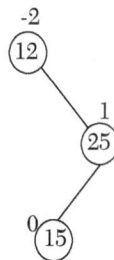

Figure 8.28(b)

After inserting 15 in the AVL tree, the balance factor of 12 is disturbed. Thus, 12 is the critical node. Hence, here we will apply an RL rotation to restore the balance factor of the tree. After rotation node 15 becomes the root node, node 12 and node 25 become the left and the right child of the tree respectively.

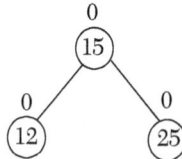

Figure 8.28. Showing an RL rotation in an AVL tree.

Therefore, the RL rotation is performed, and the balance factor of each node is also restored.

LR Rotation

LR rotation is also known as Left-Right rotation, as the new node is inserted in the right subtree of the left subtree of the critical node. It is also a double rotation. Let us take an example and perform an LR rotation in it.

For example:

Initially, the AVL tree is given as follows:

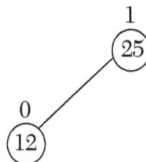

Figure 8.29(a)

Insert new node 15 in the AVL tree.

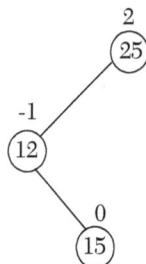

Figure 8.29(b)

After inserting 15 in the AVL tree, the balance factor of 25 is disturbed. Thus, 25 is the critical node. Hence, here we will apply an LR rotation to restore the balance factor of the tree. After rotation node 15 becomes the root node, and node 12 and node 25 become the left and the right child of the tree respectively.

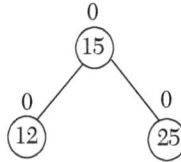

Figure 8.29. Showing an LR rotation in an AVL tree.

Therefore, an LR rotation is performed, and the balance factor of each node is also restored.

Frequently Asked Questions

2. Create an AVL tree by inserting the following elements.

60, 10, 20, 30, 19, 120, 100, 80, 19

Ans:

Step 1: *Insert 60.*

Step 2: *Insert 10. Further, no rebalancing is required.*

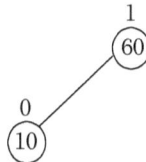

Step 3: *Insert 20. Now, rebalancing is required. We will perform LR rotation.*

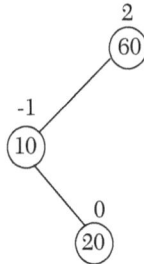

Step 4: *After performing LR rotation, the AVL tree is given as:*

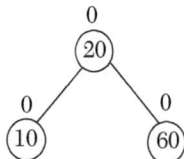

Step 5: *Insert 30. No rebalancing is required.*

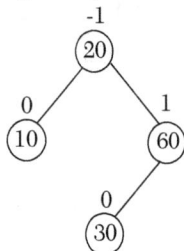

Step 6: *Insert 19. Further, no rebalancing is required.*

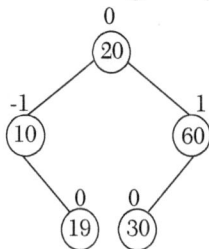

Step 7: *Insert 120. No rebalancing is required.*

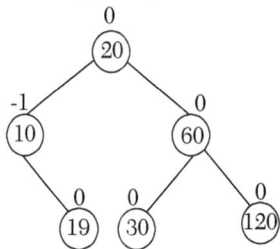

Step 8: *Insert 100. No rebalancing is required.*

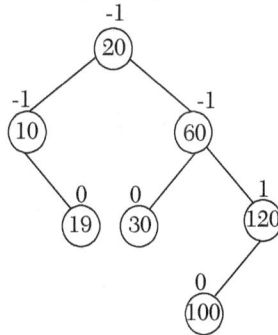

Step 9: *Insert 80. Now, rebalancing is required. We will perform LL rotation.*

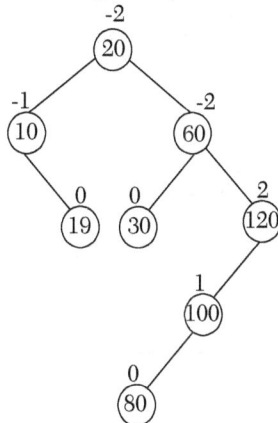

Step 10: *After performing LL rotation, the AVL tree is given as:*

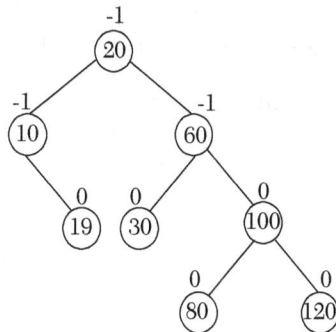

Step 11: *Insert 19. Now, rebalancing is required. We will perform RR rotation.*

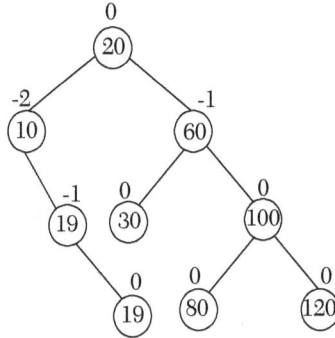

Step 12: *After performing RR rotation, the AVL tree is given as:*

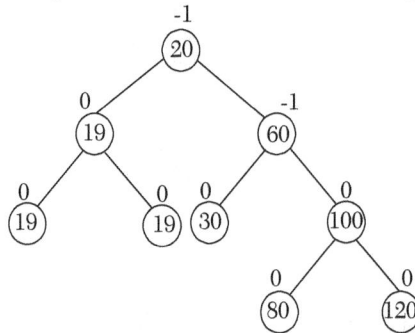

8.6   Summary

- A tree is defined as a collection of one or more nodes where one node is designated as a root node, and the remaining nodes can be partitioned into the left and the right subtrees. It is used to store hierarchical data.

- The root node is the topmost node of the tree. It does not have a parent node. If the root node is empty, then the tree is empty. A leaf node is one which does not have any child nodes.

- A path is a unique sequence of consecutive edges which is required to be followed to reach the destination from a given source.

- The degree of a node is equal to the number of children that a node has.

- A binary tree is a collection of nodes where each node contains three parts, that is, a left child, a right child, and the data item. A binary tree can have at most 2 children; that is, a parent can have either 0, 1, or 2 children.

- There are two types of binary trees, that is, complete binary trees and extended binary trees.

- In a complete binary tree, every level except the last one must be completely filled. Also, all the nodes in the complete binary tree must appear left as much as possible.

- Extended binary trees are also known as 2T-trees. A binary tree is said to be an extended binary tree if and only if every node in the binary tree has either 0 children or 2 children.

- Binary trees can be represented in the memory in two ways, which are array representation of binary trees and linked representation of binary trees. Array representation, also known as sequential representation, of binary trees is done using one-dimensional (1-D) arrays. Linked representation of binary trees is done using linked lists.

- A binary search tree (BST) is a variant of a binary tree in which all the nodes in the left subtree have a value less than that of a root node. Similarly, all the nodes in the right subtree have a value more than that of a root node. It is also known as an ordered binary tree.

- The searching operation is one of the most common operations performed in a binary search tree. This operation is performed to find whether a particular key exists in the tree or not.

- An insertion operation is performed to insert a new node with the given value in a binary search tree.

- The mirror image of a binary search tree means interchanging the right subtree with the left subtree at every node of the tree.

- Traversing is the process of visiting each node in the tree exactly once in a particular order. A tree can be traversed in various ways, which are pre-order traversal, in-order traversal, and post-order traversal.

- The word "pre" in "pre-order" determines that the root node is accessed before accessing any other node in the tree. Hence, it is also known as a DLR traversal, that is, Data Left Right.

- The word "in" in "in-order" determines that the root node is accessed in between the left and the right subtrees. Hence, it is also known as an LDR traversal, that is, Left Data Right.

- The word "post" in "post-order" determines that the root node will be accessed last after the left and the right subtrees. Hence, it is also known as an LRD traversal, that is, Left Right Data.

▪ A binary tree can be constructed if we are given at least two of the traversal results, provided that one traversal should always be an in-order traversal and the second can be either a pre-order traversal or post-order traversal.

▪ An AVL is a self-balancing binary search tree. Every node in the AVL tree has a balance factor associated with it. The balance factor is calculated by subtracting the height of the right subtree from the height of the left subtree. Thus, a node with a balance factor of -1, 0, or 1 is said to be a height-balanced tree.

8.7 Exercises

8.7.1 Theory Questions

1. What is a tree? Discuss its various applications.

2. Differentiate between height and level in a tree.

3. Explain the concept of binary trees.

4. In what ways can a binary tree be represented in the computer's memory?

5. What is a binary search tree?

6. List the various operations performed on binary search trees.

7. How can a node be deleted from a binary search tree? Discuss all the cases in detail with examples.

8. Create a binary search tree by inserting the following keys: 76, 12, 56, 31, 199, 17, 40, 76, 75. Also, find the height of the binary search tree.

9. Create a binary search tree by performing following operations:
   (a) Insert 50, 34, 23, 87, 100, 67, 43, 51, 18, and 95.
   (b) Delete 100, 34, 95 and 50 from the binary search tree.
   (c) Find the smallest value in the binary search tree.

10. How can we find the mirror image of a binary search tree?

11. List the various traversal methods of a binary tree.

12. What do you understand about an AVL tree?

13. Explain the concept of balance factor in AVL trees.

14. List the advantages of an AVL tree.

15. Consider the following binary search tree and perform the following operations:

(a) Find the pre-order and post-order traversals of the tree.

(b) Insert 25, 32, 50, 75, and 87 in the tree.

(c) Find the largest value in the tree.

(d) Delete the root node.

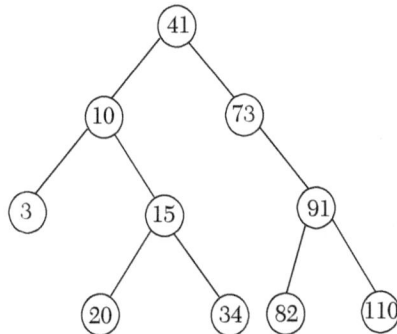

16. Give the linked representation of the previous binary search tree.

17. Construct a binary search tree of the word VIVEKANANDA. Find its pre-order, in-order, and post-order traversal.

18. Create an AVL tree by inserting the following keys, 50, 19, 59, 90, 100, 12, 10, and 150, into the tree.

19. Consider the following AVL search tree and perform various operations in it:

(a) Insert 100, 58, 93, 40, and 7 into the tree.

(b) Search for 93 in the AVL tree.

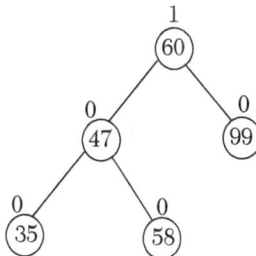

20. Discuss the various types of rotations performed in AVL trees.

21. Which one of the following is better and why? (i) AVL Trees or (ii) Binary Search Trees

22. Consider the following tree and answer the following:
(a) Determine the height of the tree.
(b) Name the leaf nodes.
(c) Siblings of C.
(d) Level number of the node J.
(e) Root node of the tree.
(f) Left and right subtrees.
(g) Depth of the tree.
(h) Ancestors of E.
(i) Descendants of H.
(j) Path from node A to F.

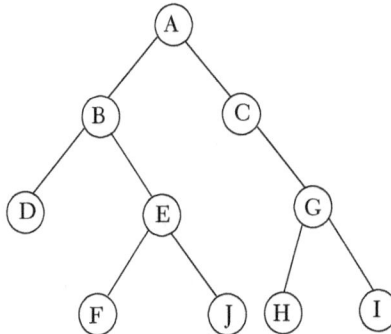

8.7.2 Programming Questions

1. Write a function to find the height of a binary search tree.

2. Write a Java program to insert and delete nodes from a binary search tree.

3. Write a Java program to show insertion in AVL trees.

4. Write a function to calculate the total number of nodes in a tree.

5. Write a Java program to traverse a binary search tree showing all the traversal methods.

6. Write a function to find the largest value in a binary search tree.

7. Write an algorithm showing post-order traversal of a binary search tree.

8. Write an algorithm to find the total number of internal nodes in a binary search tree.

9. Write a function to search for a node in a binary search tree.

8.7.3 Multiple Choice Questions

1. The maximum height of a binary tree with n number of nodes is _____.

    (a) 0

    (b) n

    (c) n+1

    (d) n-1

2. The degree of a terminal node is always _____.

    (a) 1

    (b) 2

    (c) 0

    (d) 3

3. A binary tree is a tree in which _____.

    (a) Every node must have two children

    (b) Every node must have at least two children

    (c) No node can have more than two children

    (d) All of these

4. What is the post-order traversal of the binary search tree having pre-order traversal as DBAEFGCH and in-order traversal as BEAFDCHG?

    (a) EFBAHGCD

    (b) EFBAHCGD

    (c) EFABHGCD

    (d) EFABHCGD

5. How many rotations are required during the construction of an AVL tree if the following keys are to be added in the order given?

36, 51, 39, 24, 29, 60, 79, 20, 28

    (a) 3 Left rotations, 3 Right rotations

    (b) 2 Left rotations, 2 Right rotations

    (c) 2 Left rotations, 3 Right rotations

    (d) 3 Left rotations, 2 Right rotations

6. A binary tree of height h has at least h nodes and at most _____ nodes.

   (a) 2

   (b) 2^h

   (c) $2^h - 1$

   (d) $2^h + 1$

7. How many distinct binary search trees can be created out of four distinct keys?

   (a) 5

   (b) 12

   (c) 14

   (d) 23

8. Nodes at the same level that also share same parent are called _____.

   (a) Cousins

   (b) Siblings

   (c) Ancestors

   (d) Descendants

9. The balance factor of a node is calculated by _____.

   (a) $\text{Height}_{\text{Left subtree}} - \text{Height}_{\text{Right subtree}}$

   (b) $\text{Height}_{\text{Right subtree}} - \text{Height}_{\text{Left subtree}}$

   (c) $\text{Height}_{\text{Left subtree}} + \text{Height}_{\text{Right subtree}}$

   (d) $\text{Height}_{\text{Right subtree}} + \text{Height}_{\text{Left subtree}}$

10. The following sequence is inserted into an empty binary search tree:
   6 11 26 12 5 7 16 8 35

   What is the type of traversal given by the following:
   6 5 11 7 26 8 12 35 16

   (a) Pre-order traversal

   (b) In-order traversal

   (c) Post-order traversal

   (d) None of these

11. In tree creation, which one will be the most suitable and effective data structure?
   (a) Stack
   (b) Linked list
   (d) Queue
   (d) Array

12. A binary tree can be represented as:
   (a) Linked List
   (b) Arrays
   (c) Both of the above
   (d) None of the above

13. A binary tree of n nodes has exactly n+1 edges.
   (a) True
   (b) False
   (c) Not possible to comment

14. The in-order traversal of a tree will yield a sorted listing of the elements of trees in
   (a) Binary heaps
   (b) Binary trees
   (c) Binary search trees
   (d) All of these

15. Which is the nearest ancestor node on the path from the root node to the newly inserted node of the AVL tree having balance factor -1, 0, or 1?
   (a) Parent node
   (b) Child node
   (c) Root node
   (d) Critical node

CHAPTER 9

MULTI-WAY SEARCH TREES

9.1 Introduction

We have already studied binary search trees and have discussed that every node in a binary search tree contains three parts, that is, an information part as well as a LEFT child and a RIGHT child, which store the addresses to the left and right subtrees. The same concept is used for multi-way search trees. *An M-way search tree is a tree which contains (M – 1) values per node*. It also has M subtrees. In an M-way search tree, M is called the degree of the node. For example, if the value of M = 3 in an M-way search tree, then the tree will contain two values per node, and it will have three subtrees. When an M-way search tree is not empty, it has the following properties:

1. Each node in a M-way search tree is of the following structure:

n	P_0	K_0	P_1	K_1	P_2	K_2 ------	P_{n-1}	K_{n-1}	P_n

where P_0, P_1, P_2, . . . P_n are the node's subtrees, and K_0, K_1, K_2, . . . K_n are the key values stored in the node.

2. The key values in a node are stored in ascending order, that is, $K_i < K_{i+1}$, where i = 0, 1, 2, . . . n-2.

3. All the key values stored in the left subtree are always less than the root node.

4. All the key values stored in the right subtree are always greater than the root node.

5. The subtrees pointed to by P_i for i = 0, 1, 2, . . . n are also M-way search trees.

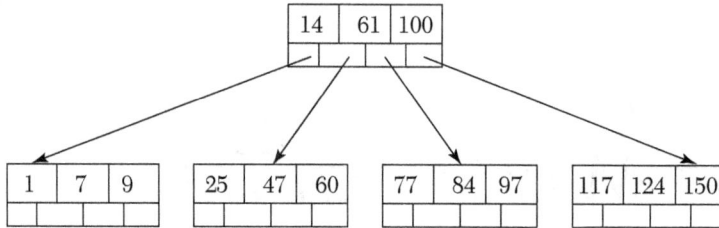

Figure 9.1. M-way search tree of order 4.

9.2   B-Trees

A B-tree is a specialized multi-way tree which is widely used for disk access. The B-tree was developed in 1970 by Rudolf Bayer and Ed McCreight. In a B-tree each node may contain a large number of keys. A B-tree is designed to store a large number of keys in a single node so that the height remains relatively small. A B-tree of order m has all the properties of a multi-way search tree. In addition, it has the following properties:

1. All leaf nodes are at the bottom level or at the same level.

2. Every node in a B-tree can have at most m children.

3. The root node can have at least two children if it is not a leaf node, and it can obviously have no children if it is a leaf node.

4. Each node in a B-tree can have at least (m/2) children except the root node and the leaf node.

5. Each leaf node must contain at least ceil [(m/2) – 1]keys.

For example: A B-tree of order 5 can have at least ceil [5/2] = 3 children and ceil [(5/2) – 1] = 2 keys. Obviously, the maximum number of children a node can have is 5. Each leaf node must contain at least 2 keys.

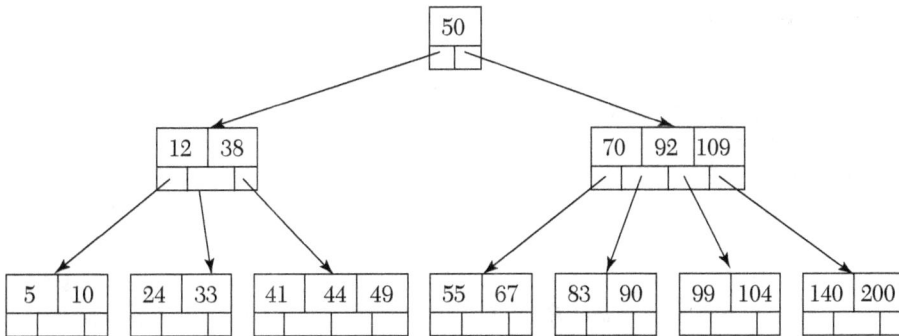

Figure 9.2. B-tree of order 4.

Practical Application

In database programs, the data is too large to fit in memory; therefore, it is stored in secondary storage, that is, tapes or disks.

9.3 Operations on a B-Tree

A B-tree stores sorted data, and we can perform on it the following operations:

- Inserting a new element in a B-tree

- Deleting an element from a B-tree

So, let's discuss both these operations in detail.

9.3.1 Insertion in a B-Tree

First of all, insertions in a B-tree are done at the leaf-node level. The following are the steps for inserting an element in a B-tree:

Step 1: In Step 1, we will search the B-tree to find the leaf node where the new key is to be inserted.

Step 2: Now, if the leaf node is full, that is, if it already contains (m – 1) keys, then follow these steps:

    i. Insert the new key into the existing set of keys in order.

    ii. Now, the node is split into two halves.

    iii. Finally, push the middle (median) element upward to its parent node. Also, if the parent node is full, then split the parent node by following these steps.

Step 3: If the leaf node is not full, that is, if it contains $(m - 1)$ keys, then insert the new key into the node, keeping the elements of the node in order.

Frequently Asked Questions

1. Construct a B-tree of order 5 and insert the following values into it:

Values to be inserted – B, N, G, A, H, E, J, Q, M, D, V, L, T, Z
Ans:
1. *Since order = 5, we can store at least 3 values and at most 4 values in a single node. Hence, we will insert B, N, G, A into the B-tree in sorted order.*

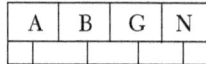

A	B	G	N

Figure 9.3(a)

2. *Now H is to be inserted between G and N, so now the order will be A B G H N, which is not possible, as at most 4 values can be accommodated in a single node. So now we will split the node, and the middle element G will become the root node.*

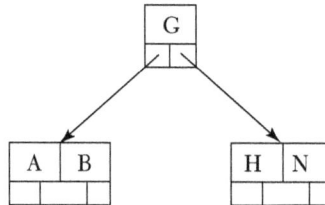

Figure 9.3(b)

3. *Now we will insert E J and Q into the B-tree.*

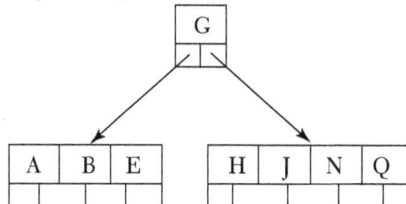

Figure 9.3(c)

4. *M is to be inserted in the right subtree. But at most 4 values can be stored in the node, so now we will push the middle element, that is, M, into the root node. Thus, the node is split into two halves.*

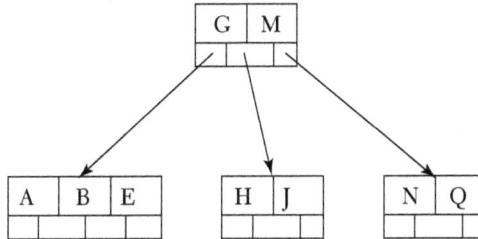

Figure 9.3(d)

5. *Now we will insert D V L T into the tree.*

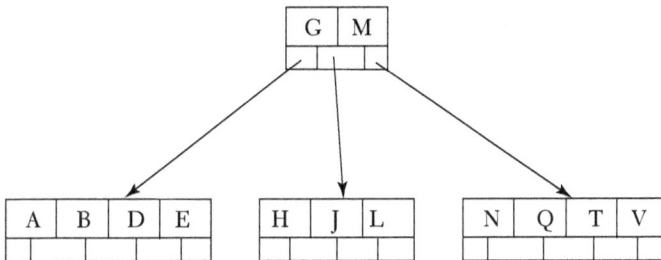

Figure 9.3(e)

6. *Finally, Z is to be inserted. It will be inserted in the right subtree. Hence, the last node will split into two halves, and the middle element, that is, T, will push up to the root node.*

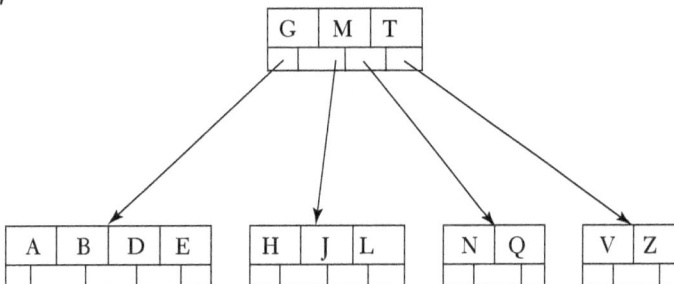

Figure 9.3(f)

9.3.2 Deletion in a B-Tree

Deletion of keys in a B-tree also first requires traversal in the B-tree; that is, after reaching a particular node, we can come across two cases which are:

1. Node is a leaf node.

2. Node is not a leaf node.

Now, let us discuss both these cases in detail.

1. Node is a leaf node

If the node has more than a minimum number of keys, then deletion can be done very easily. But if the node has a minimum number of keys, then first we will check the number of keys in the adjacent leaf node. If the number of keys in the adjacent node is greater than the minimum number of keys, then the first key of the adjacent leaf node will go to the parent node and the key present in the parent node will be combined in a single leaf node. If the parent node also has less than the minimum number of keys, then the same steps will be repeated until we get a node which has more than the minimum number of keys present in it.

2. Node is not a leaf node

In this case the key from the node is deleted, and its place will be occupied by either its successor or predecessor key. If both predecessor and successor nodes have keys less than the minimum number, then the keys of the successor and predecessor are combined.

For example: Consider a B-tree of order 5.

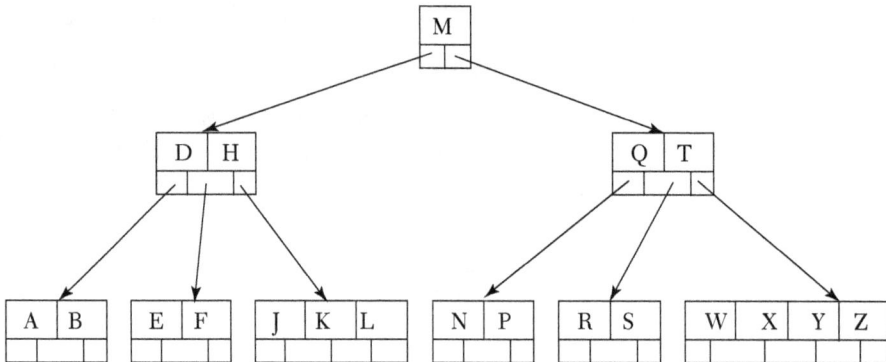

Figure 9.4(a)

1. Delete J from the tree. J is in the leaf node, so it is simply deleted from the B-tree.

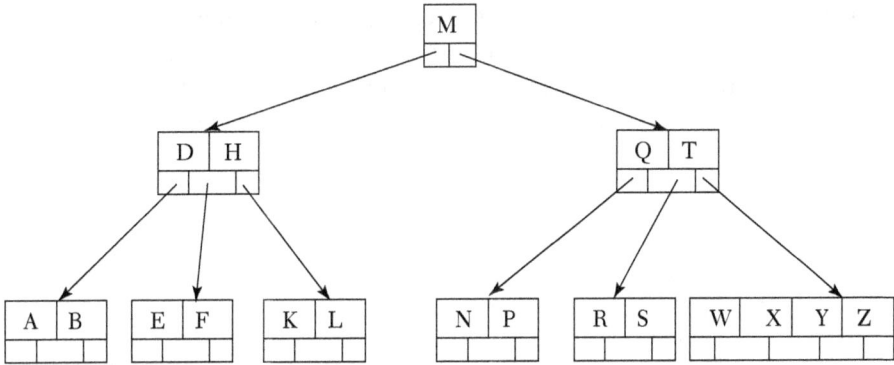

Figure 9.4(b)

2. Now T is to be deleted, but it is not in the leaf node, so we will replace T with its successor, that is, W. Hence, T is deleted.

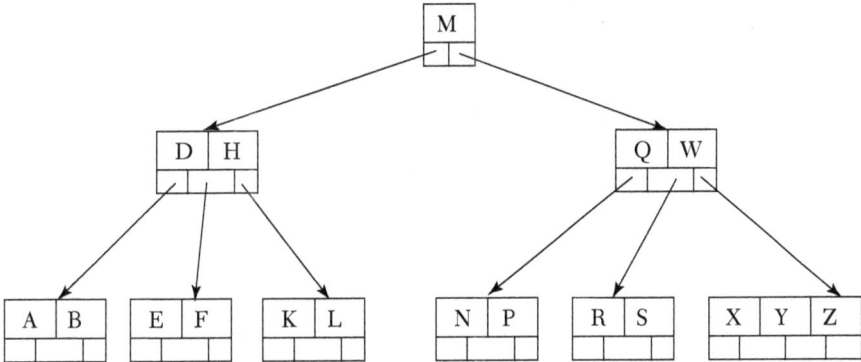

Figure 9.4(c)

3. Now delete R; in this case we will borrow keys from the adjacent leaf node.

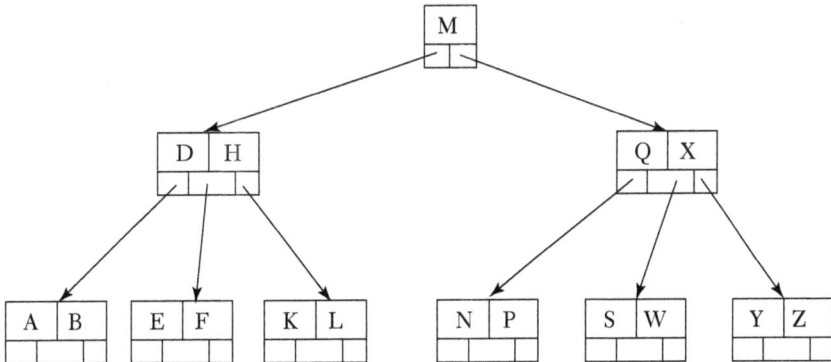

Figure 9.4(d)

4. Now we want to delete E. In this case we will also borrow keys from an adjacent node. But we can see that there are no free keys in an adjacent node, so the leaf node has to be combined with one of its two siblings. This includes moving down the parent's key that was between those two leaves.

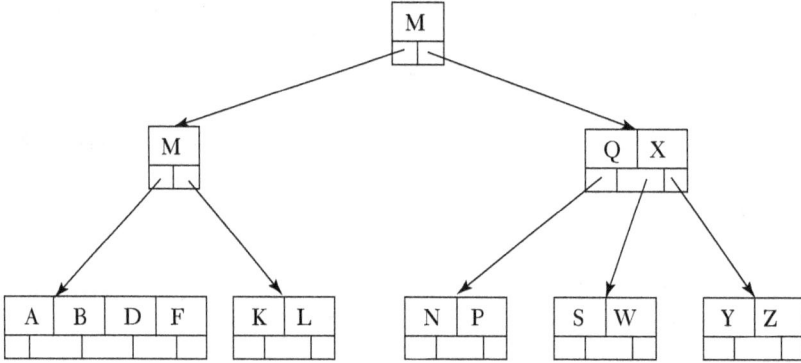

Figure 9.4(e)

But we can see that H is still unstable according to the definition. Therefore, the final tree after all deletions is shown as follows:

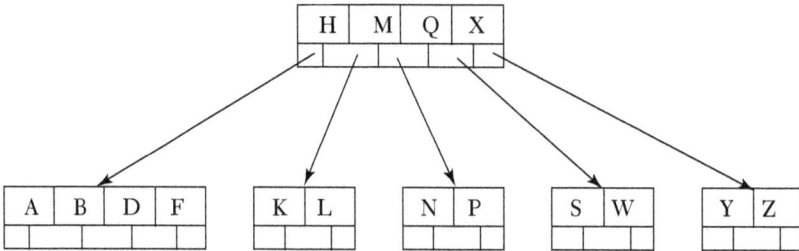

Figure 9.4(f)

Frequently Asked Questions

2. Consider the following B-tree of order 5 and insert 81, 7, 49, 61 and 30 in it.

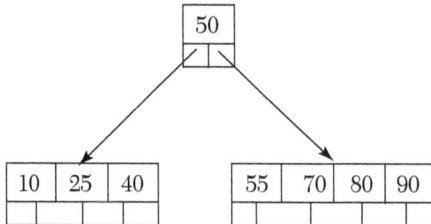

Ans:

1. Insert 81

```
              ┌────┬────┐
              │ 50 │ 80 │
              └────┴────┘
    ┌───────────┼────────────┐
┌────┬────┬────┐  ┌────┬────┐  ┌────┬────┐
│ 10 │ 25 │ 40 │  │ 55 │ 70 │  │ 81 │ 90 │
└────┴────┴────┘  └────┴────┘  └────┴────┘
```

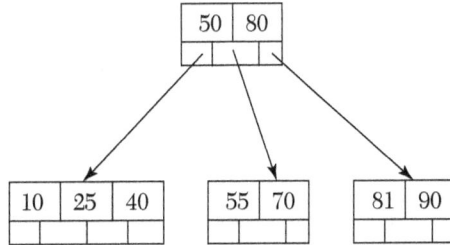

2. Insert 7 and 49

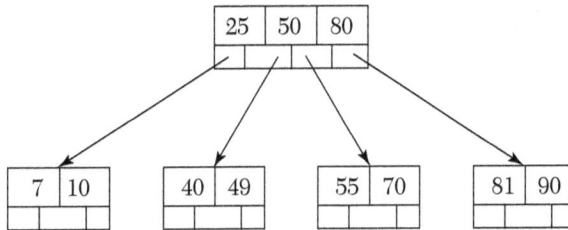

```
              ┌────┬────┬────┐
              │ 25 │ 50 │ 80 │
              └────┴────┴────┘
    ┌──────────┼──────┼──────────┐
┌────┬────┐ ┌────┬────┐ ┌────┬────┐ ┌────┬────┐
│ 7  │ 10 │ │ 40 │ 49 │ │ 55 │ 70 │ │ 81 │ 90 │
└────┴────┘ └────┴────┘ └────┴────┘ └────┴────┘
```

3. Insert 61 and 30

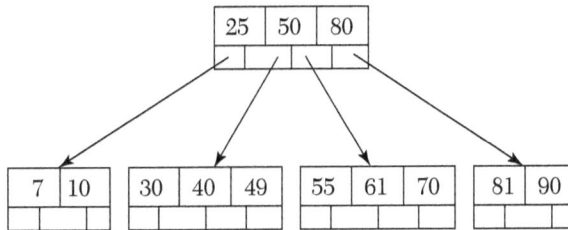

```
              ┌────┬────┬────┐
              │ 25 │ 50 │ 80 │
              └────┴────┴────┘
    ┌──────────┼──────┼──────────┐
┌────┬────┐ ┌────┬────┬────┐ ┌────┬────┬────┐ ┌────┬────┐
│ 7  │ 10 │ │ 30 │ 40 │ 49 │ │ 55 │ 61 │ 70 │ │ 81 │ 90 │
└────┴────┘ └────┴────┴────┘ └────┴────┴────┘ └────┴────┘
```

Figure 9.5. Insertion in a B-tree.

Frequently Asked Questions

3. Consider the following B-tree of order 5 and delete the values 95, 200, 176, and 70 from it.

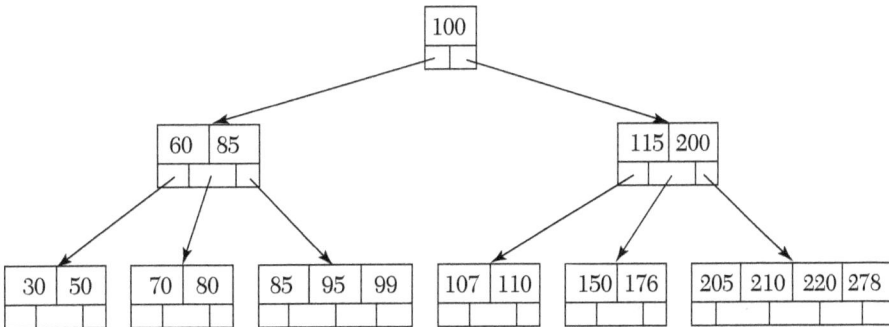

```
                          ┌─────┐
                          │ 100 │
                          └─────┘
          ┌─────────────────┴─────────────────────┐
    ┌────┬────┐                              ┌─────┬─────┐
    │ 60 │ 85 │                              │ 115 │ 200 │
    └────┴────┘                              └─────┴─────┘
  ┌────┬────┬────┐                    ┌────────┼──────────┐
┌────┬────┐┌────┬────┐┌────┬────┬────┐┌─────┬─────┐┌─────┬─────┐┌─────┬─────┬─────┬─────┐
│ 30 │ 50 ││ 70 │ 80 ││ 85 │ 95 │ 99 ││ 107 │ 110 ││ 150 │ 176 ││ 205 │ 210 │ 220 │ 278 │
└────┴────┘└────┴────┘└────┴────┴────┘└─────┴─────┘└─────┴─────┘└─────┴─────┴─────┴─────┘
```

Ans:
1. Delete 95

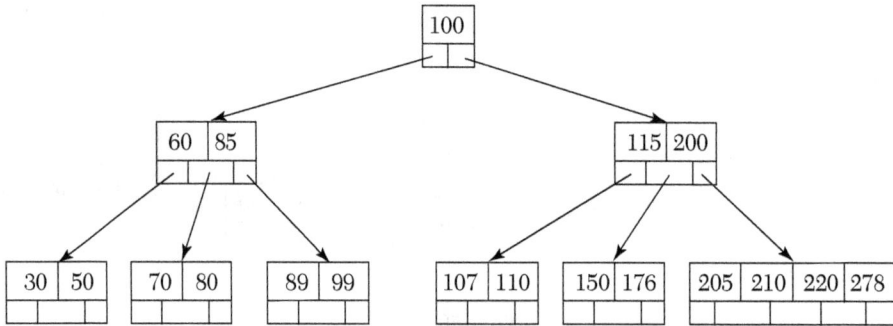

2. Delete 200

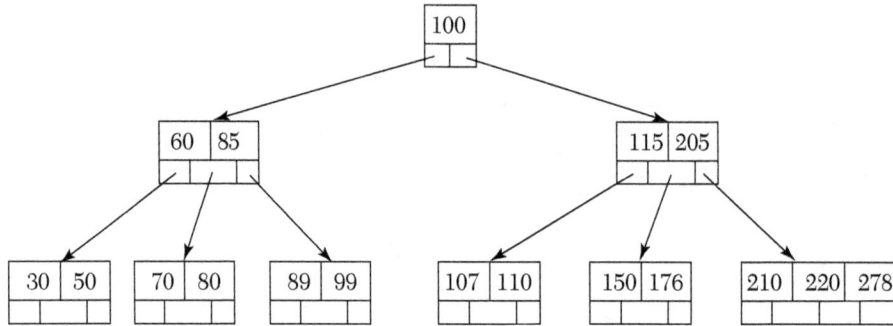

3. Delete 176

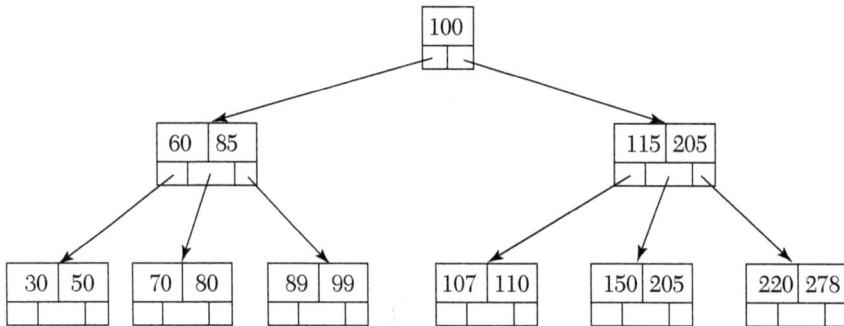

4. Delete 70

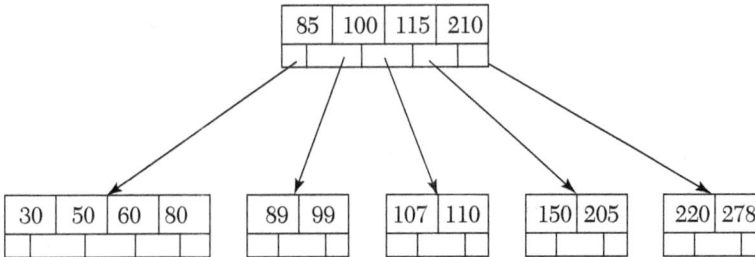

Figure 9.6. Deletion in a B-tree.

9.4 Application of a B-Tree

The main application of a B-tree is the organization of a large amount of data or a huge collection of records into a file structure. A B-tree should search the records very efficiently, and all the operations such as insertion, deletion, searching, and so on should be done very efficiently; therefore, the organization of records should be very good.

9.5 B+ Trees

A B+ tree is a variant of a B-tree which also stores sorted data like a B-tree. The structure of aB-tree is the standard organization for indexes in database systems. Multilevel indexing is done in a B+ tree; that is, leaf nodes constitute a dense index, while non-leaf nodes constitute a sparse index. A B+ tree is a slightly different data structure which allows sequential processing of data and stores all the data in the lowest level of the tree. A B-tree can store both records and keys in its interior nodes, while a B+ tree stores all the records in its leaf nodes and the keys in its interior nodes. In a B+ tree, the leaf nodes are linked to one another like a linked list. A B+ tree is usually used to store big amounts of data which cannot be stored in the primary memory. Hence, in a B+ tree the leaf nodes are stored in the secondary storage, while the internal nodes are stored in the main memory.

In a B+ tree, all the internal nodes are called index nodes because they store the index values. Similarly, *all the external nodes are called data nodes* because they store the keys. A B+ tree is always balanced and is very efficient for the searching of data, as all the data is stored in the leaf nodes. Various advantages of a B+ tree are as follows:

(a) A B+ tree is always balanced, and the height of the tree always remains less as compared to other tree structures.

(b) All the leaf nodes are linked to one another, which make it very efficient.

(c) The leaf nodes are also linked to the nodes at an upper level; thus, it can be easily used for a wide range of search queries.

(d) The records can be fetched in equal number of disk access.

(e) The records can be accessed either sequentially or randomly.

(f) Searching of data becomes very simple, as all the information is stored only in leaf nodes.

(g) Similarly, deletion is also very simple, as it will only take place in the leaf nodes.

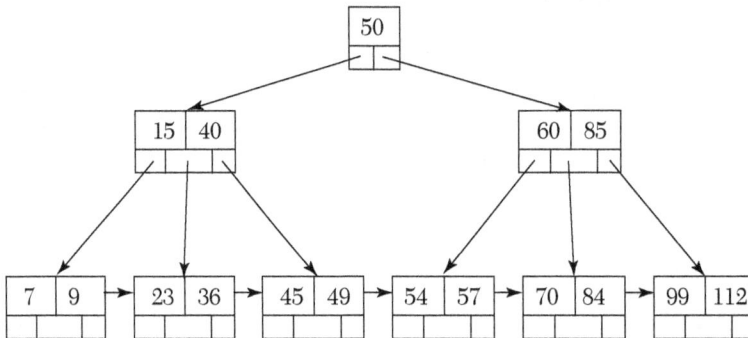

Figure 9.7. B+ tree of order 3.

9.6 Summary

- An M-way search tree has M – 1 values per node and M subtrees. M is called the degree of the node.

- A B-tree is a specialized multi-way tree which is widely used for disk access. The B-tree was developed in 1970 by Rudolf Bayer and Ed McCreight.

- A B-tree of order m has all the properties of a multi-way search tree.

- The main application of a B-tree is the organization of a large amount of data or a huge collection of records into a file structure.

- A B+ tree is a variant of a B-tree which also stores sorted data like a B-tree. The structure of a B-tree is the standard organization for indexes in database systems. A B+ tree is a slightly different data structure which allows sequential processing of data and stores all the data in the lowest level of the tree.

9.7   Exercises

9.7.1   Review Questions

1. Define
   (a) M-way search tree
   (b) B-tree
   (c) B+ tree

2. Write a difference between B-trees and B+ trees.

3. Construct a B-tree of order 3, inserting the keys 10, 20, 50, 60 40, 80, 100, 70, 130, 90, 30, 120, 140, 25, 35, 160, 180 in a left-to-right sequence. Show the trees on deleting 190 and 60.

4. Explain the insertion and deletion of a node in a B-tree.

5. Explain B+ tree indexing with the help of an example.

6. What do you mean by B-trees? Write the steps to create a B-tree. Construct an M-way search tree of order 4 and insert the values 34, 45, 98, 1, 23, 41, 78, 100, 234, 122, 199, 10, 40.

7. Why do we always prefer a higher value of m in a B-tree? Explain.

8. Are B-trees of order 2 full binary trees? Explain.

9.7.2   Multiple Choice Questions

1. B+ trees are preferred to binary trees in databases because:
   (a) Disk capacities are greater than memory capacities.
   (b) Disk access is slower than memory access.
   (c) Disk data transfer rates are less than the memory data transfer rates.
   (d) Disks are more reliable than memory.

2. In an M-way search tree, M stands for _____.
   (a) Degree of the node
   (b) External nodes
   (c) Internal nodes
   (d) None of these

3. A B-tree of order 4 is built. What is the maximum number of keys that a node may accommodate before splitting operations take place?
   (a) 5

    (b) 2

    (c) 4

    (d) 3

4. In a B-tree of order m, every node has at the most _____ children.

    (a) M + 1

    (b) M – 1

    (c) M/2

    (d) M

5. Which is the best data structure to search the keys in less time?

    (a) B-tree

    (b) M-way search tree

    (c) B+ tree

    (d) Binary search tree

6. The best case of searching a value in a binary search tree is:

    (a) $O(n^2)$

    (b) $O(\log n)$

    (c) $O(n)$

    (d) $O(n \log n)$

7. External nodes are also called _____.

    (a) Index nodes

    (b) Data nodes

    (c) Value nodes

    (d) None of the above

8. AB+ tree stores redundant keys.

    (a) False

    (b) True

    (c) Not possible to comment

9. A B-tree of order 5 can store at least how many keys?

    (a) 0

    (b) 1

    (c) 2

    (d) 3

CHAPTER 10

HASHING

10.1   Introduction

In Chapter 6, we discussed three types of searching techniques: linear search, binary search, and interpolation search. Linear search has a running time complexity of O(n), whereas binary search has a running time proportional to O(log n), where n is equal to the number of elements in the array. The searching algorithms discussed within Chapter 6 are efficient. However, their search time is dependent on the number of elements in the array, and none of them can search for an element within the constant time equal to O(1).But it is very difficult to achieve in all the searching algorithms like linear search, binary search, and so on, as all these algorithms are dependent on the number of elements present in the array. Also, there are many comparisons involved while searching for an element using the previous searching algorithms. Therefore, our primary need is to search for the element in a constant time along with fewer key comparisons. Now, let us take an example. Suppose there is an array of size N and all the keys to be stored in the array are unique and also are in the range 0 to N-1. Now, we will store all the records in the array based on the key where the array index numbers and keys are the same. Thus, in that case we can access the records in a constant time along with no key comparisons involved in it. This can be further explained by the following figure:

arr[0]	arr[1]	arr[2]	arr[3]	arr[4]	arr[5]	arr[6]	arr[7]	arr[8]	arr[9]
	1		3	4		6		8	

Figure 10.1. An array.

In the previous figure, there is an array containing five elements. Note that the keys and the array index numbers are the same; that is, the record with the key value 3 can be directly accessed by array index arr[3]. Similarly, all the records can be accessed through key values and the array index. Thus, this can be done by hashing, where we will convert the key into an array index and store the records in the array. This can be done as follows:

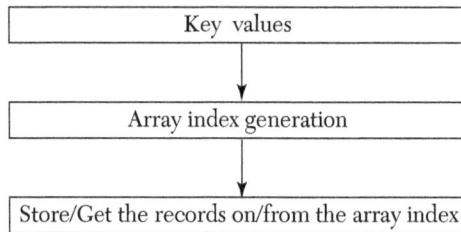

Figure 10.2. Array index generation using hashing.

The process of array index generation uses a hash function which is used to convert the keys into an array index. The array in which such records are stored is known as a hash table.

Practical Application

1. A simple real-life example is when we search for a word in the dictionary and then find the definition or meaning with the help of a key and its index.

2. Driver's license numbers and insurance card numbers are created using hashing from data items that never change, that is, date of birth, name, and so on.

Frequently Asked Questions

1. Explain the term hashing.

Ans: *Hashing is the process of mapping keys to their appropriate locations in the hash table. It is the most effective technique of searching for the values in an array or in a hash table.*

10.1.1 Difference between Hashing and Direct Addressing

In direct addressing, we store the key at the same address as the value of the key as shown in Figure 10.3. However, in hashing, as shown in Figure 10.4, the address of the key is determined by using a mathematical function known as a hash function. The hash function will operate on the key to determine the address of the key. Direct addressing may result in a more random distribution of the key throughout the memory, and hence sometimes leads to more wastage of space when compared with hashing.

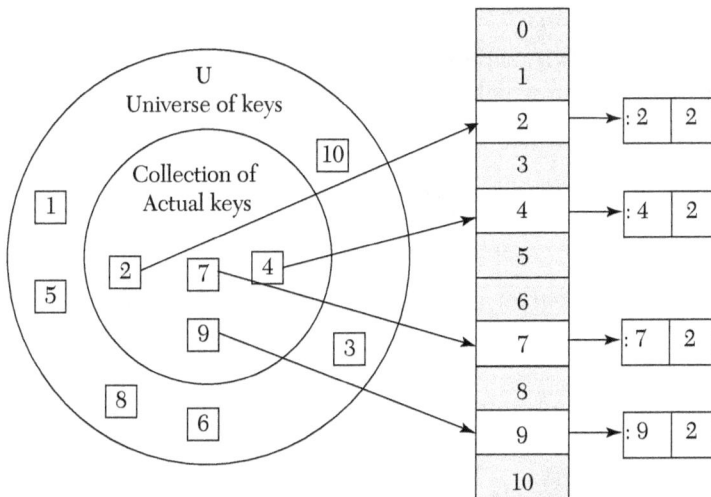

Figure 10.3. Mapping of keys using a direct addressing method.

10.1.2 Hash Tables

A hash table is a data structure which supports one of the efficient searching techniques, that is, hashing. A hash table is an array in which the data is accessed through a special index called a key. In a hash table, keys are mapped to the array positions by a hash function. A hash function is a function or mathematical formula which, when applied to a key, produces an integer which is used as an index to find a key in the hash table. Thus, a value stored in a hash table can be searched in $O(1)$ time with the help of a hash function. The main idea behind a hash table is to establish direct mapping between the keys and the indices of the array.

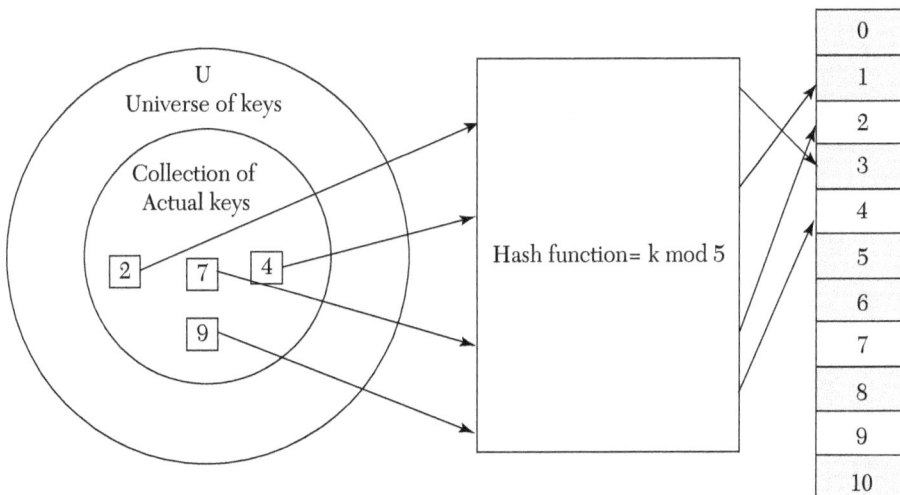

Figure 10.4. Mapping of keys to the hash table using hashing.

10.1.3 Hash Functions

A hash function is a mathematical formula which, when applied to a key, produces an integer which is used as an index to find a key in the hash table.

Characteristics of the Hash Function

There are four main characteristics of hash functions which are as follows:

1. The hash function uses all the input data.

2. The hash function must generate different hash values.

3. The hash value is fully determined by the data being hashed.

4. The hash function must distribute the keys uniformly across the entire hash table.

Different Types of Hash Functions

In this section, we will discuss some of the common hash functions:

1. Division Method: In the division method, a key k is mapped into one of the m slots by taking the remainder of k divided by m. In simple terms, we can say that this method divides an integer, say x, by m and then uses the remainder so obtained. It is the simplest method of hashing. The hash function is given by:

$$h(k) = k \ mod \ m$$

Address   Key   m.no. of slots

For example, if m = 5 and the key k = 10, then h(k) = 2. Thus, the division method works very fast, as it requires only a single division operation. Although this method is good for any value of m, consider that if m is an even number then h(k) is even when the value of k is even, and similarly h(k) is odd when the value of k is odd. Therefore, if the even and odd keys are almost equal, then there will be no problem. But if there is a larger number of even keys, then the division method is not good, as it will not distribute the keys uniformly in the hash table. Also, we avoid certain values of m; that is, m should not be a power of 2, because if h(k) = k mod 2^x, then h(k) will extract the lowest x bits of k. The main drawback of the division method is that many consecutive keys map to consecutive hash values, which means that consecutive array locations will be occupied, and hence there will be an effect on the performance.

Frequently Asked Questions

2. Given a hash table of 50 memory locations, calculate the hash values of keys 20 and 75 using the division method.

Ans:
m = 50, k1 = 10, k2 = 75 hash values are calculated as:
h(10) = 10 % 50 = 10
h(75) = 75 % 50 = 25

2. **Mid Square Method:** In the mid square method, we will calculate the square of the given key. After getting the number, we will extract some digits from the middle of that number as an address.

   For example, if key k = 5025, then k^2 = 25250625. Thus, h(5025) = 50. This method works very well, as all the digits of the key contribute to the output; that is, all the digits contribute in producing the middle digits. In addition, the same r digits must be chosen from all the keys in this method.

Frequently Asked Questions

3. Given a hash table of 100 memory locations, calculate the hash values of keys 2045 and 1357 using the mid square method.

Ans: *Now, there are 100 memory locations where indices will be from 0 to 99. Hence, only two digits will be taken to map the keys. So, the value of r is equal to 2.*
k_1 = 2045, k2 = 4182025, h(2045) = 20
k_2 = 1357, k2= 1841449, h(1357) = 14
Note: The third and fourth digits are chosen to start from the right.

3. **Folding Method:** In the folding method, we will break the key into pieces such that each piece has the same number of digits except the last one, which may have fewer digits as compared to the other pieces. Now, these individual pieces are added. We will ignore the carry if it exists. Hence, the hash value is formed.

   For example, if m = 100 and the key k = 12345678, then the indices will vary from 0 to 99, and thus each piece of the key must have two digits. Therefore, the given key will be broken into four pieces, that is, 12, 34, 56, and 78. Now we will add all these, that is, 12 + 34 + 56 + 78 = 180. Thus, the hash value will be 80 (ignore the carry).

Frequently Asked Questions

4. Given a hash table of 100 memory locations, calculate the hash values of keys 2486 and 179 using the folding method.

Ans: *Now, there are 100 memory locations where indices will be from 0 to 99. Hence, each piece of the key must have two digits.*
$h(2486) = 24 + 86 = 110$
$h(2486) = 10$ *(ignore the last carry)*
$h(179) = 17 + 9 = 26$
$h(179) = 26$

10.1.4   Collision

A collision is a situation which occurs when a hash function maps two different keys to a single/same location in the hash table. Suppose we want to store a record at one location. Now, another record cannot be stored at the same location as it is obvious that two records cannot be stored at the same location. Thus, there are methods to solve this problem, which are called collision resolution techniques.

10.1.5   Collision Resolution Techniques

As already discussed, collision resolution techniques are used to overcome the problem of collision in hashing. There are two popular methods which are used for resolving collisions:

1. Collision Resolution by Chaining Method

2. Collision Resolution by Open Addressing Method

Now, we will discuss these methods in detail.

10.1.5.1   Chaining Method

In the chaining method, a chain of elements is maintained which have the same hash address. The hash table here behaves like an array of references. Each location in the hash table stores a reference to the linked list, which contains all the key elements that were hashed to that location. For example, location 5 in the hash table points to the key values that hashed to location 5. If no key value hashes to location 5, then in that case location 5 will contain NULL. The following figure shows how the key values are mapped to the hash table and also how they are stored in the linked list.

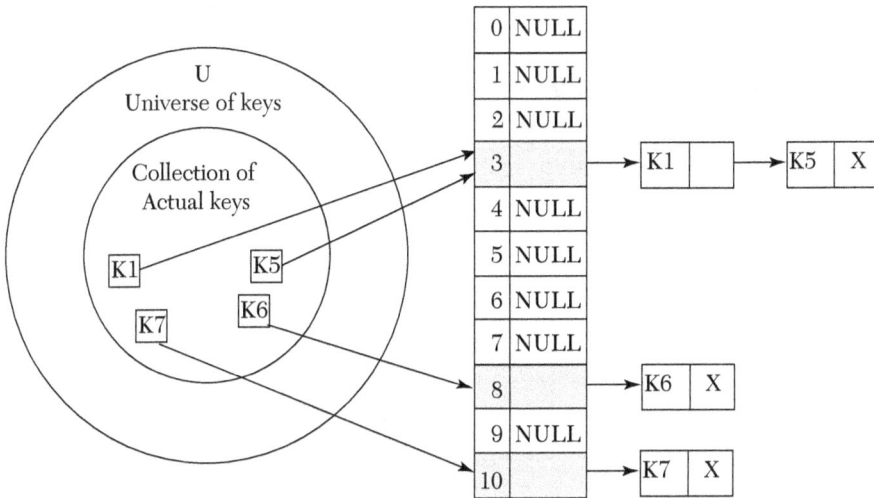

Figure 10.5. Keys being hashed by chaining method.

Operations on a Chained Hash Table

1. **Insertion in a Chained Hash Table:** The process of inserting an element is quite simple. First, we get the hash value from the hash function, which will map to the hash table. After mapping, the element is inserted in the linked list. The running time complexity of inserting an element in a chained hash table is O(1).

2. **Deletion from a Chained Hash Table:** The process of deleting an element from a chained hash table is the same as we used in the singly linked list. First, we will perform a search operation, and then the delete operation as in the case of the singly linked list is performed. The running time complexity of deleting an element from a chained hash table is O(m), where m is the number of elements present in the linked list at that location.

3. **Searching in a Chained Hash Table:** The process of searching for an element in a chained hash table is also very simple. First, we will get the hash value of the key by the hash function in the hash table. Then we will search for the element in the linked list. The running time complexity of searching for an element in a chained hash table is O(m), where m is the number of elements present in the linked list at that location.

Frequently Asked Questions

5. Insert the keys 4, 9, 20, 35, and 49 in a chained hash table of 10 memory locations. Use hash function h(k) = k mod m.

Ans: *Initially, the hash table is given as:*

0	NULL
1	NULL
2	NULL
3	NULL
4	NULL
5	NULL
6	NULL
7	NULL
8	NULL
9	NULL

Now, we will insert 4 in the hash table.

Step 1:

Key to be inserted = 4

h(4) = 4 mod 10

h(4) = 4

Now, we will create a linked list for location 4, and the key element 4 is stored in it.

0	NULL
1	NULL
2	NULL
3	NULL
4	→ 4 X
5	NULL
6	NULL
7	NULL
8	NULL
9	NULL

Step 2:
Key to be inserted = 9
$h(9) = 9 \bmod 10$
$h(9) = 9$
Now, we will create a linked list for location 9, and the key element 9 is stored in it.

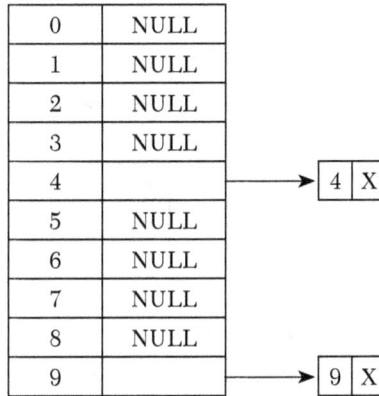

0	NULL
1	NULL
2	NULL
3	NULL
4	→ 4 X
5	NULL
6	NULL
7	NULL
8	NULL
9	→ 9 X

Step 3:
Key to be inserted = 20
$h(20) = 20 \bmod 10$
$h(20) = 2$
Now, we will create a linked list for location 2, and the key element 20 is stored in it.

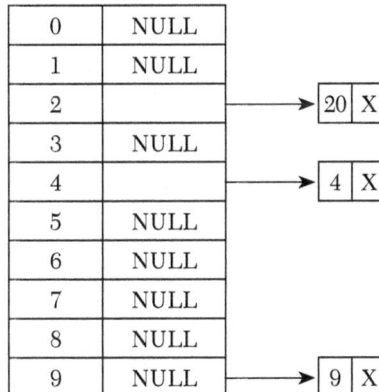

0	NULL
1	NULL
2	→ 20 X
3	NULL
4	→ 4 X
5	NULL
6	NULL
7	NULL
8	NULL
9	NULL → 9 X

Step 4:
Key to be inserted = 35
$h(35) = 35 \bmod 10$
$h(35) = 5$
Now, we will create a linked list for location 5, and the key element 35 is stored in it.

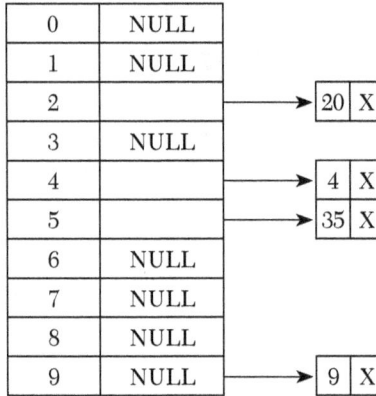

0	NULL
1	NULL
2	→ 20 X
3	NULL
4	→ 4 X
5	→ 35 X
6	NULL
7	NULL
8	NULL
9	NULL → 9 X

Step 5:
Key to be inserted = 49
$h(49) = 49 \bmod 10$
$h(49) = 9$
Now, we will insert 49 at the end of the linked list of location 9.

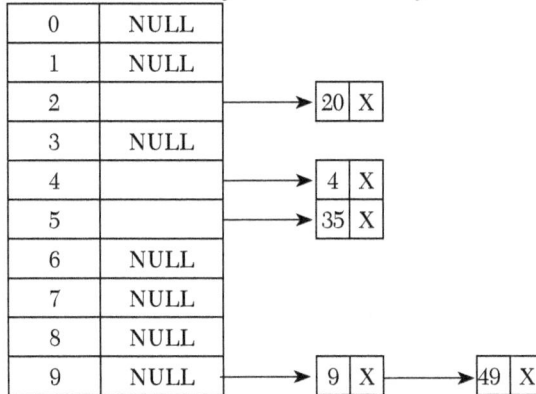

0	NULL
1	NULL
2	→ 20 X
3	NULL
4	→ 4 X
5	→ 35 X
6	NULL
7	NULL
8	NULL
9	NULL → 9 X → 49 X

Advantages and Disadvantages of the Chained Method

The main advantage of this method is that it completely resolves the problem of collision. It remains effective even when the key elements to be stored in the hash table are higher than the number of locations in the hash table. However, it is quite obvious that with the increase in the number of key elements, the performance of this method will decrease.

The disadvantage of this method is the wastage of storage space as the key elements are stored in the linked list;in addition,the referencesare required for each element to get accessed, which in turn are consuming more space.

10.1.5.2 Open Addressing Method

In the open addressing method, all the elements are stored in the hash table itself. Once a collision takes place, open addressing computes new locations using the probe sequence, and the next element or next record is stored on that location. Probing is the process of examining the memory locations in the hash table. When we perform the insertion operation in the open addressing method, we first successively probe/examine the hash table until we find an empty slot in which the new key can be inserted. The open addressing method can be implemented using the following:

- Linear Probing

- Quadratic Probing

- Double Hashing

Now, let us discuss all of them in detail.

Linear Probing

Linear probing is the simplest approach to resolving the problem of collision in hashing. In this method, if a key is already stored at a location generated by the hash function h(k), then the situation can be resolved by the following hash function:

$$h\ (k) = (h(k) + i) = mod\ m$$

where $h(k) = k\ mod\ m$

i = Probe no. = 0, 1, 2, 3(m - 1)

m = no. of slots

Now, let us understand the working of this technique. For a given key k, first the location generated by (h(k) + 0) mod m is probed, because for the first time i = 0. If the location generated is free, then the key is stored in it. Otherwise, the second probe is generated for i = 1 given by the hash function (h(k) + 1) mod m. Similarly, if the location generated is free, then the key is stored in it; otherwise, subsequent probes are generated such as (h(k) + 2) mod m, (h(k) + 3) mod m, and so on, until we find a free location.

Frequently Asked Questions

6. Given keys k = 13, 25, 14, and 35, map these keys into a hash table of size m = 5 using linear probing.

Ans: *Initially, the hash table is given as:*

0	1	2	3	4
NULL	NULL	NULL	NULL	NULL

Step 1:
i = 0
Key to be inserted = 13
h'(k) = (k mod m + i) mod m
h'(13) = (13 % 5 + 0) % 5
h'(13) = (3 + 0) % 5
h'(13) = 3 % 5 = 3
Now, since location T[3] is free, 13 is inserted at location T[3].

0	1	2	3	4
NULL	NULL	NULL	13	NULL

Step 2:
i = 0
Key to be inserted = 25
h'(25) = (25 % 5 + 0) % 5
h'(25) = (0 + 0) % 5
h'(13) = 0 % 5 = 0
Now, since location T[0] is free, 25 is inserted at location T[0].

0	1	2	3	4
25	NULL	NULL	13	NULL

Step 3:
i = 0
Key to be inserted = 14
h'(14) = (14 % 5 + 0) % 5
h'(14) = (4 + 0) % 5
h'(14) = 4 % 5 = 4
Now, since location T[4] is free, 14 is inserted at location T[4].

0	1	2	3	4
25	NULL	NULL	13	14

Step 4:
i = 0
Key to be inserted = 35
h'(35) = (35 % 5 + 0) % 5
h'(35) = (0 + 0) % 5
h'(35) = 0 % 5 = 0
Now, since location T[0] is not free, the next probe sequence,that is, i = 1, is
computed as:
i = 1
h'(35) = (35 % 5 + 1) % 5
h'(35) = (0 + 1) % 5
h'(35) = 1 % 5 = 1
Now, since location T[1] is free, 35 is inserted at location T[1].
Thus, the final hash table is shown as:

	0	1	2	3	4
	25	35	NULL	13	14

// Write a program to show the linear probing technique of the collision resolution method.

```java
import java.util.Scanner;
public class LinearProbing {
    static int SIZE = 10;
    static int[] arr = new int[SIZE];
    public static void main(String[] args) {
        Scanner scn = new Scanner(System.in);
        boolean flag = true;
        for (int i = 0; i < SIZE; i++) {
            arr[i] = 0;
            while (flag == true) {
                System.out.println("\nMENU");
                System.out.println("1. Insert Keys");
                System.out.println("2. Search Keys");
                System.out.println("3. Display Keys");
                System.out.println("4. Exit ");
                System.out.println("\nSelect Operation: ");
                int ch = scn.nextInt();
                switch (ch) {
                case 1:
                    System.out.println("Enter values of key: ");
                    int k1 = scn.nextInt();
                    if (k1 != -1) {
```

```
                        insertion(k1, arr);
                    }
            System.out.println("Inserted Successfully!!!");
                    break;
            case 2:
                System.out.println("Enter key value to
                                    search: ");
                int k2 = scn.nextInt();
                lprob(k2, arr);
                break;
            case 3:
                display(arr);
                break;
            case 4:
                System.out.println("Terminated...");
                flag = false;
                break;
            default:
                System.out.println("Wrong Choice");
            }
        }
    }
}
public static void insertion(int k, int arr[]) {
    int position;
    position = k % SIZE;
    while (arr[position] != 0) {
        position = ++position % SIZE;
    }
    arr[position] = k;
}
public static void lprob(int k, int arr[]) {
    int position;
        position = k % SIZE;
    while ((arr[position] != k) && (arr[position] != 0)) {
        position = ++position % SIZE;
    }
    if (arr[position] != 0)
        System.out.println("Successfully searched at: " +
                            position);
    else
        System.out.println("Unsuccessfull search");
}
public static void display(int arr[]) {
    int i;
```

```
        System.out.println(" List of keys :\n");
    for (i = 0; i < SIZE; i++)
        System.out.println("\t" + arr[i]);
    }
}
```

The output of the program is shown as:

Advantages and Disadvantages of Linear Probing

Linear probing is a very good technique, as the algorithm provides good memory caching through good locality of address. But the main disadvantage of this method is that it results in clustering. Due to clustering, there is a higher risk of collisions taking place. Also, the time required for searching also increases with the size of the clusters. Now, we can say that the higher the number of collisions, the higher the number of probes required to find a vacant location, and the performance is lessened. This is known as primary clustering. We can avoid this clustering by using other techniques such as quadratic probing and double hashing.

Quadratic Probing

Quadratic probing is another approach to resolving the problem of collision in hashing. In this method, if a key is already stored at a location generated by the hash function h(k), then the situation can be resolved by the following hash function:

$$h (k) = (h(k) + c1i = c2i^2) \bmod m$$

where $h(k) = k \bmod m$

i = Probe no. = 0, 1, 2, 3...$(m - 1)$

$c1, c2$ = consonants

$(c1, c2$ should not be equal to zero)

The quadratic probing method is better than linear probing, as it terminates the phenomenon of primary clustering because of its searching speed; that is, it is doing a quadratic search. For a given key k, first the location generated by (h(k) + 0 + 0) mod m is probed, because for the first time i = 0. If the location generated is free, then the key is stored in it. Otherwise, subsequent positions probed are offset by the amounts/factors that depend in a quadratic manner on the probe number i. The quadratic probing method works better than linear probing, but to maximize the use of the hash table, the values of m, c_1, and c_2 are constrained.

Frequently Asked Questions

7. Given keys k = 25, 13, 14, and 35, map these keys into a hash table of size m = 5 using quadratic probing with c_1 = 1 and c_2 = 3.

Ans: *Initially, the hash table is given as follows:*

0	1	2	3	4
NULL	NULL	NULL	NULL	NULL

Step 1:

$i = 0$

$c_1 = 1, c_2 = 3$

Key to be inserted = 25

$h'(k) = (k \bmod m + c_1 i + c_2 i2) \bmod m$

$h'(25) = (25 \% 5 + 1 \times 0 + 3 \times (0)2) \% 5$

$h'(25) = (0 + 0) \% 5$

$h'(13) = 0 \% 5 = 0$

Now, since location T[0] is free, 25 is inserted at location T[0].

0	1	2	3	4
25	NULL	NULL	NULL	NULL

Step 2:

$i = 0$

$c_1 = 1, c_2 = 3$

Key to be inserted = 13

$h'(13) = (13 \% 5 + 1 \times 0 + 3 \times (0)2) \% 5$

$h'(13) = (3 + 0) \% 5$

$h'(13) = 3 \% 5 = 3$

Now, since location T[3] is free, 13 is inserted at location T[3].

0	1	2	3	4
25	NULL	NULL	13	NULL

Step 3:

$i = 0$

$c_1 = 1, c_2 = 3$

Key to be inserted = 14

$h'(14) = (14 \% 5 + 1 \times 0 + 3 \times (0)2) \% 5$

$h'(14) = (4 + 0) \% 5$

$h'(14) = 4 \% 5 = 4$

Now, since location T[4] is free, 14 is inserted at location T[4].

0	1	2	3	4
25	NULL	NULL	13	14

Step 4:

$i = 0$

$c_1 = 1, c_2 = 3$

Key to be inserted = 35

$h'(35) = (35 \% 5 + 1 \times 0 + 3 \times (0)2) \% 5$

$h'(35) = (0 + 0) \% 5$

$h'(35) = 0 \% 5 = 0$

Now, since location T[0] is not free, the next probe sequence,that is, i = 1, is computed as:

$i = 1$

$h'(35) = (35 \% 5 + 1 \times 1 + 3 \times (1)2) \% 5$

$h'(35) = (0 + 1 + 3) \% 5$

$h'(35) = 4 \% 5 = 4$

Again, since location T[4] is not free, the next probe sequence,that is, i = 2, is computed as:

$i = 2$

$h'(35) = (35 \% 5 + 1 \times 2 + 3 \times (2)2) \% 5$

$h'(35) = (0 + 2 + 12) \% 5$

$h'(35) = 14 \% 5 = 4$

Again, since location T[4] is not free, the next probe sequence,that is, i = 3, is computed as:

$i = 3$

$h'(35) = (35 \% 5 + 1 \times 3 + 3 \times (3)2) \% 5$

$h'(35) = (0 + 3 + 27) \% 5$

$h'(35) = 30 \% 5 = 0$

Again, since location T[0] is not free, the next probe sequence,that is, i = 4, is computed as:

$i = 4$

$h'(35) = (35 \% 5 + 1 \times 4 + 3 \times (4)2) \% 5$

$h'(35) = (0 + 4 + 48) \% 5$

$h'(35) = 52 \% 5 = 2$

Now, since location T[2] is free, 35 is inserted at location T[2].

Thus, the final hash table is shown as:

0	1	2	3	4
25	NULL	35	13	14

// Write a program to show the quadratic probing technique of the collision resolution method.

```java
import java.util.Scanner;
public class QuadraticProbing {
    static int SIZE = 10;
    static int[] arr = new int[SIZE];
    public static void main(String[] args) {
        Scanner scn = new Scanner(System.in);
        boolean flag = true;
        for (int i = 0; i < SIZE; i++) {
            arr[i] = 0;
            while (flag == true) {
                System.out.println("\nMENU");
                System.out.println("1. Insert Keys");
                System.out.println("2. Search Keys");
                System.out.println("3. Display Keys");
                System.out.println("4. Exit ");
                System.out.println("\nSelect Operation: ");
                int ch = scn.nextInt();
                switch (ch) {
                case 1:
                    System.out.println("Enter values of
                                        key: ");
                    int k1 = scn.nextInt();
                    if (k1 != -1) {
                        insertion(k1, arr);
                    }
                    System.out.println("Inserted
                                        Successfully!!!");
                    break;
                case 2:
                    System.out.println("Enter key value to
                                        search: ");
                    int k2 = scn.nextInt();
                    qprob(k2, arr);
                    break;
                case 3:
                    display(arr);
                    break;
                case 4:
                    System.out.println("Terminated...");
                    flag = false;
                    break;
```

```
                    default:
                        System.out.println("Wrong Choice");
                    }
                }
            }
        }
    public static void insertion(int k, int arr[]) {
        int position;
        position = k % SIZE;
        while (arr[position] != 0) {
            position = ++position % SIZE;
        }
        arr[position] = k;
    }
    public static void qprob(int k, int arr[]) {
        int position;
        position = k % SIZE;
        while ((arr[position] != k) && (arr[position] != -1)) {
            position = ++position % SIZE;
        }
        if (arr[position] != 0)
            System.out.println("Successfully searched at: " +
                                position);
        else
            System.out.println("Unsuccessfull search");
    }
    public static void display(int arr[]) {
        int i;
        System.out.println(" List of keys :\n");
        for (i = 0; i < SIZE; i++)
            System.out.println("\t" + arr[i]);
    }
}
```

The output of the program is shown as:

```
Administrator: Command Prompt                                    —  □  ×

C:\Program Files\Java\jdk-12.0.2\bin>javac QuadraticProbing.java

C:\Program Files\Java\jdk-12.0.2\bin>java QuadraticProbing
MENU
1. Insert Keys
2. Search Keys
3. Display Keys
4. Exit

Select Operation:
1
Enter values of key:
25
Inserted Successfully!!!

MENU
1. Insert Keys
2. Search Keys
3. Display Keys
4. Exit

Select Operation:
1
Enter values of key:
12
Inserted Successfully!!!

MENU
1. Insert Keys
2. Search Keys
3. Display Keys
4. Exit

Select Operation:
1
Enter values of key:
56
Inserted Successfully!!!

MENU
1. Insert Keys
2. Search Keys
3. Display Keys
4. Exit

Select Operation:
1
Enter values of key:
58
Inserted Successfully!!!
MENU
1. Insert Keys
2. Search Keys
3. Display Keys
4. Exit

Select Operation:
2
Enter key value to search:
56
Successfully searched at: 6
MENU
1. Insert Keys
2. Search Keys
3. Display Keys
4. Exit

Select Operation:
3
List of keys :

          0
          0
          12
          0
          0
          25

          56
          0
          58
          0

MENU
1. Insert Keys
2. Search Keys
3. Display Keys
4. Exit

Select Operation:
4
Terminated...

C:\Program Files\Java\jdk-12.0.2\bin>
```

Advantages and Disadvantages of Quadratic Probing

As previously discussed, one of the biggest advantages of quadratic probing is that it eliminates the phenomenon of primary clustering. Yet one of the major disadvantages of this method is that a sequence of successive probes may only cover some portion of the hash table, and this portion may be quite small. Therefore, if such a situation occurs, then it will be difficult for us to find an empty location in the hash table, despite the fact that the table is not full. Hence, quadratic probing encounters a problem which is known as secondary clustering. In this method, the chance of multiple collisions increases as the hash table becomes full. This type of situation can be overcome by double hashing.

Double Hashing

Double hashing is one of the best methods available for open addressing. As the name suggests, this method uses two hash functions to operate rather than a single hash function. The hash function is given as follows:

$$h'(k) = (h_1(k) + ih_2(k)) \bmod m,$$

where $h_1(k) = k \bmod m$ and $h_2(k) = k \bmod m'$ are the two hash functions, m is the size of the hash table, m' is less than m (can be $(m-1)$ or $(m-2)$), and i is the probe number that varies from 0 to $(m-1)$.

Now, let us understand the working of this technique. For a given key k, first the location generated by $(h_1(k) \bmod m)$ is probed, because for the first time $i = 0$. If the location generated is free, then the key is stored in it. Otherwise, subsequent probes generate locations that are at an offset of $(h_2(k) \bmod m)$ from the previous location. Also, the offset may vary with every probe depending upon the value generated by the second hash function, that is, $(h_2(k) \bmod m)$. As a result, the performance of double hashing is very near to the performance of the "ideal" scheme of uniform hashing.

Frequently Asked Questions

8. Given keys k = 71, 29, 38, 61, and 100, map these keys into a hash table of size m = 5 using double hashing. Take h_1 = (k mod 5) and h_2 = (k mod 4).

Ans:
Initially, the hash table is given as:

	0	1	2	3	4
	NULL	NULL	NULL	NULL	NULL

Step 1:
i = 0
Key to be inserted = 71
$h'(k) = (h_1(k) + ih_2(k)) \bmod m$
$h'(k) = (k \bmod m + (i\, k \bmod m')) \bmod m$

$h'(71) = (71 \% 5 + (0 \times 71 \% 4)) \% 5$
$h'(71) = (1 + (0 \times 3)) \% 5$
$h'(71) = 1 \% 5 = 1$
Now, since location T[1] is free, 71 is inserted at location T[1].

0	1	2	3	4
NULL	71	NULL	NULL	NULL

Step 2:
i = 0
Key to be inserted = 29
$h'(k) = (k \bmod m + (i\,k \bmod m')) \bmod m$
$h'(29) = (29 \% 5 + (0 \times 29 \% 4)) \% 5$
$h'(29) = (4 + (0 \times 1)) \% 5$
$h'(29) = 4 \% 5 = 4$
Now, since location T[4] is free, 29 is inserted at location T[4].

0	1	2	3	4
NULL	71	NULL	NULL	29

Step 3:
i = 0
Key to be inserted = 38
$h'(k) = (k \bmod m + (i\,k \bmod m')) \bmod m$
$h'(38) = (38 \% 5 + (0 \times 38 \% 4)) \% 5$
$h'(38) = (3 + (0 \times 2)) \% 5$
$h'(38) = 3 \% 5 = 3$
Now, since location T[3] is free, 38 is inserted at location T[3].

0	1	2	3	4
NULL	71	NULL	38	29

Step 4:
i = 0
Key to be inserted = 61
$h'(k) = (k \bmod m + (i\,k \bmod m')) \bmod m$
$h'(61) = (61 \% 5 + (0 \times 61 \% 4)) \% 5$
$h'(61) = (1 + (0 \times 1)) \% 5$
$h'(61) = 1 \% 5 = 1$
Now, since location T[1] is not free, the next probe sequence, that is, i = 1, is computed as:
i = 1
$h'(61) = (61 \% 5 + (1 \times 61 \% 4)) \% 5$

$h'(61) = (1 + (1 \times 1)) \% 5$
$h'(61) = (1 + 1) \% 5$
$h'(61) = 2 \% 5 = 2$
Now, since location T[2] is free, 61 is inserted at location T[2].

0	1	2	3	4
NULL	71	61	38	29

Step 5:
$i = 0$
Key to be inserted = 100
$h'(k) = (k \bmod m + (i\,k \bmod m')) \bmod m$
$h'(100) = (100 \% 5 + (0 \times 100 \% 4)) \% 5$
$h'(100) = (0 + (0 \times 0)) \% 5$
$h'(100) = 0 \% 5 = 0$
Now, since location T[0] is free, 100 is inserted at location T[0].
Thus, the final hash table is shown as:

0	1	2	3	4
100	71	61	38	29

Advantages and Disadvantages of Double Hashing

The double hashing method is free from all the problems of primary clustering and secondary clustering. It also minimizes repeated collisions.

10.2 Summary

- A hash table is an array in which the data is accessed through a special index called a key. In a hash table, keys are mapped to the array positions by a hash function.

- A hash function is a mathematical formula which when applied to a key produces an integer which is used as an index to find a key in the hash table.

- There are different types of hash functions which use numeric keys. Popular methods are the division method, the mid square method, and the folding method.

- In the division method, a key k is mapped into one of the m slots by taking the remainder of k divided by m. The main drawback of the division method is that many consecutive keys map to consecutive hash values respectively, which means that consecutive array locations will be occupied, and hence there will be an effect on the performance.

■ In the mid square method, we will calculate the square of the given key. After getting the number, we will extract some digits from the middle of that number as an address.

■ In the folding method, we will break the key into pieces such that each piece has the same number of digits except the last one, which may have lower digits as compared to other pieces. Now, these individual pieces are added. Hence, the hash value is formed.

■ A collision is a situation which occurs when a hash function maps two different keys to a single/same location in the hash table.

■ Collision resolution techniques are used to overcome the problem of collision in hashing. There are two popular methods thatare used for resolving collisions, which are collision resolution by the chaining methodand collision resolution by the open addressing method.

■ In the chaining method, a chain of elements is maintained which have the same hash address. Each location in the hash table stores an address to the linked list which contains all the key elements that were hashed to that location. The disadvantage of this method is the wastage of storage space as the key elements are stored in the linked list; in addition, addressesare required for each element to get accessed, which in turn consume more space.

■ In an open addressing method, all the elements are stored in the hash table itself. There is no need to provide the addressin this method, which is the biggest advantage of this method. Once a collision takes place, open addressing computes new locations using the probe sequence, and the next element or next record is stored in that location.

■ Probing is the process of examining the memory locations in the hash table.

■ Linear probing is the simplest approach to resolving the problem of collision in hashing. In this method, if a key is already stored at a location generated by the hash function h(k), then the situation can be resolved by the following hash function:

$$h'(k) = (h(k) + i) \bmod m$$

▪ Quadratic probing is another approach to resolving the problem of collision in hashing. In this method, if a key is already stored at a location generated by the hash function h(k), then the situation can be resolved by the following hash function:

$$h'(k) = (h(k) + c_1 i + c_2 i^2) \bmod m$$

▪ Double hashing is one of the best methods available for open addressing. As the name suggests, this method uses two hash functions to operate rather than a single hash function. The hash function is given as:

$$h'(k) = (h_1(k) + i h_2(k)) \bmod m$$

10.3 Exercises

10.3.1 Review Questions

1. What are hash tables?

2. What is hashing? Give some of its Practical Applications.

3. Define the hash function and also explain the various characteristics of a hash function.

4. What is a collision in hashing and how it can be resolved?

5. Explain the different types of hash functions along with examples.

6. Discuss the collision resolution techniques in hashing.

7. What is clustering in hashing? What are the two types of clustering?

8. What do you understand about double hashing?

9. Define the following terms:
   (a) Quadratic Probing
   (b) Linear Probing

10. What is the chaining method in hashing and how it can help in resolving collisions?

11. Consider a hash table of size 10. Using linear probing, insert the keys 12, 45, 67, 122, 78, and 34 in it.

12. Consider a hash table of size 9. Using double hashing, insert the keys 4, 17, 30, 55, 90, 11, 54, and 77 in it. Take $h_1 = k \bmod 9$ and $h_2 = k \bmod 6$.

13. Consider a hash table of size 11. Using quadratic probing, insert the keys 10, 45, 56, 97, 123, and 1 in it.

14. How can the open addressing method be used in resolving collisions?

15. Write a Java function to retrieve an item from the hash table using linear probing and quadratic probing.

10.3.2 Multiple Choice Questions

1. Which of the following collision resolution techniques is free from the clustering phenomenon?
  (a) Linear Probing
  (b) Quadratic Probing
  (c) Double Hashing
  (d) None of these

2. The process of examining a memory location is called _____.
  (a) Probing
  (b) Hashing
  (c) Chaining
  (d) Addressing

3. A hash table with chaining as a collision resolution technique degenerates to a:
  (a) Tree
  (b) Graph
  (c) Array
  (d) Linked List

4. Which of the probing techniques suffers from the problem of primary clustering?
  (a) Quadratic Probing
  (b) Linear Probing
  (c) Double Hashing
  (d) All of these

5. Given the hash function $h(k) = k \bmod 6$, what is the number of collisions to store the following sequence of keys, 16, 20, 45, 68, using open addressing?

   (a) 1

   (b) 3

   (c) 2

   (d) 5

6. In a hash table, an element with the key k is stored at _____.

   (a) k

   (b) $h(k^2)$

   (c) $h(k)$

   (d) $\log h(k)$

7. A good hash function eliminates the problem of collision.

   (a) True

   (b) False

   (c) Not possible to comment

8. Given the hash function of size 7 and hash function $h(k) = k \bmod 7$, what is the number of collisions with linear probing for insertion of the following keys: 29, 36, 16, and 30?

   (a) 1

   (b) 2

   (c) 3

   (d) 4

9. _____ is the process of mapping keys to appropriate locations in the hash table.

   (a) Probing

   (b) Hashing

   (c) Collision

   (d) Addressing

CHAPTER 11

FILES

11.1   Introduction

We all know that nowadays in most organizations, a large amount of data is collected in one form or another. Some of the organizations use various types of data collection applications for collecting the data. When we talk about an organization, it is not only the big ones like schools, colleges, and companies, but also a small bakery at the corner of the street; it can be observed that collection and exchange of data take place everywhere. For example, when we get admitted intoa school, a lot of data is collected by the school such as name, age, address, parent's name, blood type, and so on. We all know that in the past data was collected in the form of paper documents which were very difficult to handle and store. Therefore, to efficiently and effectively analyze the collected data, computers are used to store the data in the form of files. A file in computer terminology is defined as a block of useful data in a persistent storage medium;that is, the file is available for future use. The data is organized in a hierarchical order in the files. The hierarchical order includes items such as records, fields, and so forth, which all are defined as follows.

11.2   Terminologies

- **Data Field:** A data field is a unit which stores a unary fact. It is usually characterized by its type and size. For example, "employee's name" is a data field that stores the names of employees.

- **Record:** The collection of related data fields is called a record. For example, anemployee's record may contain various data fields such as name, ID, address, contact number, and so on.

- **File:** The collection of related records is called a file. An example is a file of the employees working in an organization.

- **Directory:** The collection of related files is called a directory. Every file in a computer system is stored in a directory.

- **File Name:** The nameof a file is a string of characters.

- **Read-only:** A file named read-only cannot be modified or deleted. If we try to delete the file, then a particular message is displayed.

- **Hidden:** A file marked as hidden is not displayed in the directory.

11.3   File Operations

There are various operations which can be performed on the files.

1. **File Creation:** It is the first operation to be performed on the files if the file has not been created. A file is created by specifying its name and mode. The records are inserted into the file by opening the file in writing mode. Once all the records are inserted into the file, the file can be used for future read and write operations. For example, we create a new file named EMPLOYEE.

    (a) **Updating a File:** It means changing the contents of a file. It is usually done in the following ways:
    (b) **Inserting into a File:** The new record is inserted into the file. For example, if a new employee joins an organization, his/her record is inserted in the EMPLOYEE file.
    (c) **Modifying a File:** The existing records are modified in the file. For example, if the address of an employee is changed, then the new address must be modified in the EMPLOYEE file.
    (d) **Deleting from a File:** The existing record is deleted from the file. For example, if an employee quits a job, then his/her record is deleted from the EMPLOYEE file.

2. **Retrieving from a File:** It refers to the process of extracting some useful data from a file. It is usually done in the following ways:
    (a) **Enquiring:** It retrieves a low amount of data from the file.
    (b) **Generating a Report:** It retrieves a huge amount of data from the file.

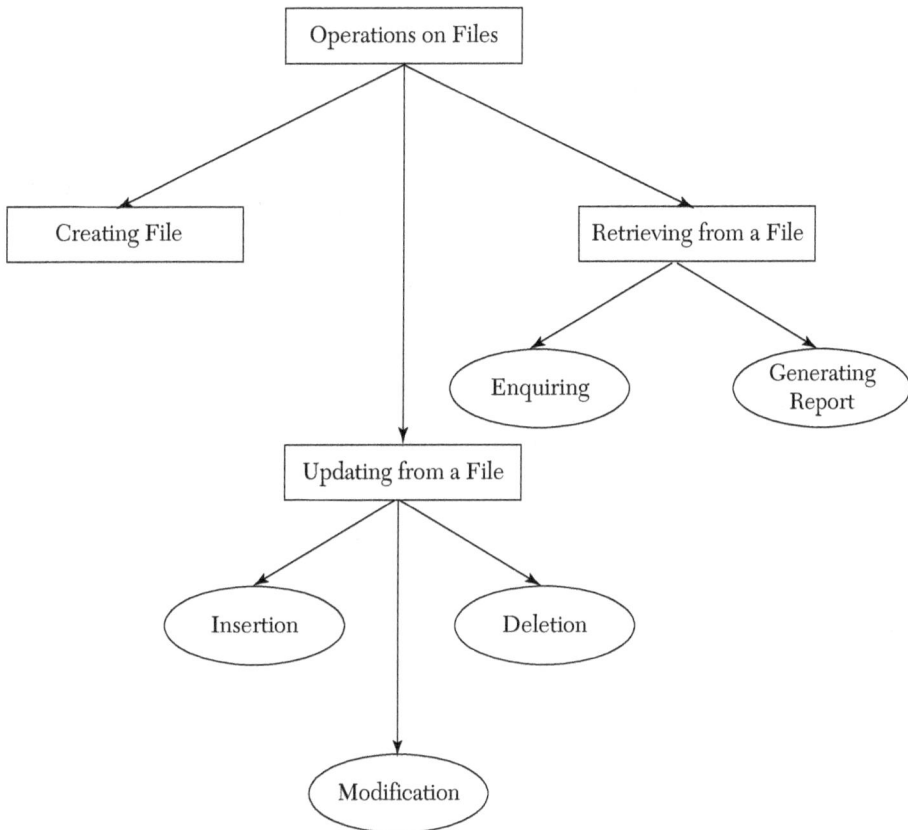

Figure 11.1. Operations on files.

11.4 File Classification

A file is classified into two types, which are:

1. **Text Files:** A text file, often called a flat file, is a file that stores all the numeric or non-numeric data using its corresponding ASCII values. The data can be a string of letters, numbers, or special symbols. Therefore, it is also known as an ASCII file. Usually, a text file has a special marker known as an end of file marker, which denotes the end of the file.

2. **Binary Files:** A binary file is a file that contains all the data in the binary form of 1s and 0s. It stores the data in the same form as that of primary memory. Thus, a binary file is not readable by human beings. Binary files are read by computer programs, and they decode the binary files into something meaningful. Data is efficiently stored in binary files.

11.5   C vs C++ vs Java File Handling

File handling is an important process, and one must be aware of the file handing process irrespective of the particular language. But especially when it comes to C, C++, and Java file handling, it becomes a little bit tough to understand the operations and processes on files, as these languages possess similar kinds of functions/operators. Hence, there are some main points to remember while working with files, which are discussed as follows:

C	C++	Java
In C, fopen, fclose, fwrite, fread, fseek, fprint, fscanf, and various other functions are called directly without any help of an object.	In C++, open, close, and other functions are called with the help of an object, for example, fstream f. Here, f is the object of the stream class, and all the functions will be called with the help of objects like f.open, f.close, f.read, f.write, and so on.	In Java, the package java.io is imported and the object of the class File, FileReader, or FileWriter is created. The objects of the classes access the inbuilt constructors and/or methods of the classes through objects.
In C, the modes are r(read), w(write), and a(append), and these can be used directly.	In C++, the modes are in, out, bin, and so on, and these are used with the help of scope resolution operators like ios::in, ios::out, and so forth.	Java uses the concept of a Stream, that is, a sequence of data.

11.6   File Organization

File organization refers to the way in which records are physically arranged on a storage device. Further, there may be a single key or multiple keys associated with it. Therefore, based on its physical storage and the keys used to access the records, files are classified as sequential files, relative files, indexed sequential files, and inverted files. There are various factors which should be taken into consideration while choosing a particular type of file organization, which are:

1. Ease of retrieval of the records.

2. Economy of storage.

3. Reliability, that is, whether a file organization is reliable or not.

4. Security, that is, whether a file organization is secured or not.

Now, we will discuss some of the techniques which are commonly used for file organization.

11.7 Sequence File Organization

Sequence file organization is the most basic way to organize a collection of records in a file. Sequence file organization is when the file is created when the records are written, one after the other in order, and can be accessed only in the order in which they are written when the file is used for input. All the records are numbered from zero onward. Thus, if there are N records in a file, then the first record is numbered as 0, and the last record will be numbered as N-1. In some cases, records of sequential files are sorted by the value of some field in each record. The field whose value is used to sort the records is known as a sort key. If a file is sorted by the value of a field named "key field," then the record i proceeds record j if and only if the value of "key field" in record i is less than or equal to the value of "key field" in record j. Also, a file can be sorted in either ascending or descending order by a sort key comprising one or more fields. As the records in a sequential file can only be accessed sequentially, these files are used more commonly in batch processing than in interactive processing. For example, the records of a sequential file are used to generate the white pages of a telephone directory that will be sorted by the subscriber's last name.

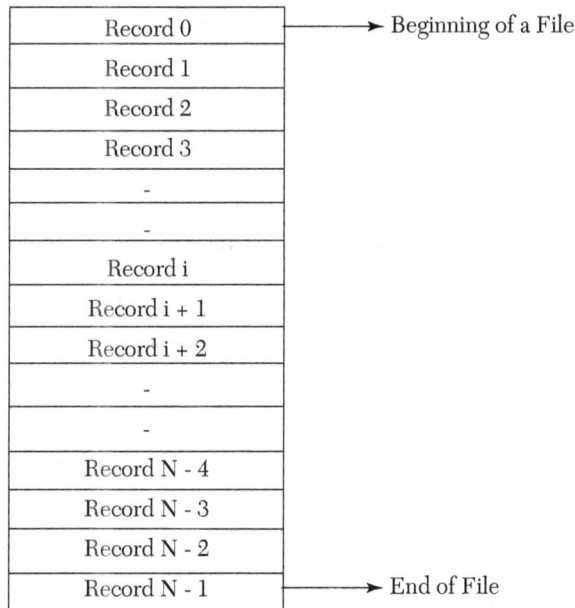

Figure 11.2. Structure of a sequence file organization.

Advantages of a Sequence File Organization

1. It is easy to handle.

2. It does not involve extra overheads/problems.

3. Records can be of varying lengths in this organization.

4. It can be stored on magnetic disks as well as tapes.

Disadvantages of Sequence File Organization

1. Records can be accessed only in sequence.

2. It does not support the update operation in between the files.

3. It does not support interactive applications.

11.8 Indexed Sequential File Organization

An indexed sequential file organization is an efficient way of organizing the records when there is a need to access both sequentially by some key values and also individually by the same key value. It provides the combination of access types that are supported by a sequential file or a relative file. The index has been structured as a binary search tree. This index is used to serve as a request for access to a particular record, and the sequential data file alone is used to support sequential access to the entire collection of records. Because of its capability to support both sequential and direct access, indexed sequential file organization is used to support applications that require both batch and interactive processing.

Advantages of Indexed Sequential File Organization

1. Records can be accessed sequentially and randomly.

2. It supports batch as well as interactively oriented applications.

3. It supports the update operation in between records in the file.

Disadvantages of Indexed Sequential File Organization

1. In this organization, files can only be stored on magnetic disks.

2. It involves extra overhead in the form of maintenance.

3. Records can only be of a fixed length, as we maintain the structure of each node like a linked list.

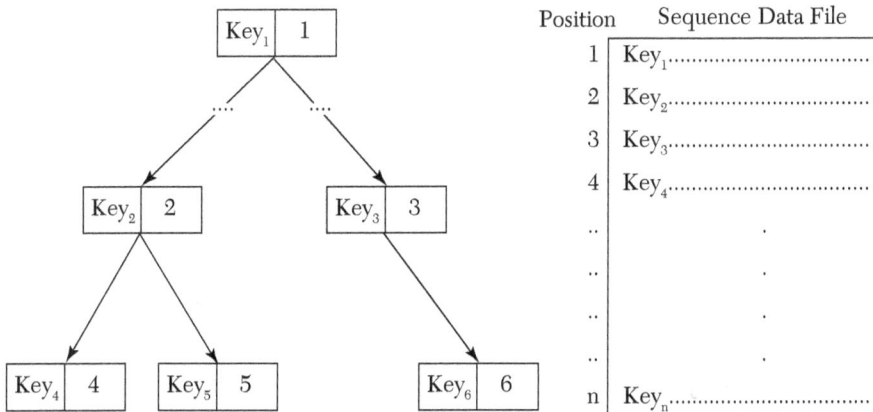

Figure 11.3. Use of BST and sequential files to provide indexed sequential access.

11.9 Relative File Organization

Relative file organization provides an effective way of accessing individual records directly. In relative file organization, there is a predictable relationship between the key and the record's location in the file. The records do not necessarily appear physically in sorted order by their keys. Then how is a given record found? The relationship that will be used to translate between key value and the physical address is designated, for example, R(Key value à address). When a record is to be written into a relative file, the mapping function R is used to translate the record's key to an address, which indicates where the record is to be stored. The fundamental techniques that are used for mapping function R are directory lookup and address calculation (hashing).

- **Directory Lookup Technique:** It is the simplest technique for implementing a mapping function R. The basic idea of this technique is to keep a directory of key values: address pairs. To find a record in a relative file, one locates its key value in the directory, and then the indicated address is used to find the record on the storage device. The directory can be organized as a binary search tree.

- **Address Calculation Technique:** Another common technique for implementing a mapping function R is to perform a calculation on the key value (hashing) such that the result is a relative address.

Advantages of Relative File Organization

1. Records can be accessed out of sequence.

2. It is well suited for interactive applications.

3. It supports an update operation in between the files.

Disadvantages of Relative File Organization

1. It can be stored only on magnetic disks.

2. It also involves extra overhead in the form of maintenance of indexes.

11.10   Inverted File Organization

One fundamental approach for providing a linkage between an index and a file is called inversion. A key's inversion index contains all the values that the key presently has in records of the file. Each key-value entry in the inversion index points to all the data records that have the corresponding value. Then, the file is said to be inverted on that key. The inversion approach for providing multi-key access has been used as the basis for a physical data structure in commercially available relational DBMS such as Oracle, DB2, and so on. These systems were designed to provide rapid access to the records via as many inversion keys as the designer cares to identify. They have user-friendly, natural-language-like query languages to assist the user in formulating inquiries. A complete inverted file has an inversion index for every data field. If a file is not completely inverted but has at least one inversion index, then it is said to be a partially inverted file.

Advantages of Inverted File Organization

1. The Boolean query requires only one access per record satisfying the query along with some access to process the indexes.

2. Records can be stored in any way, for example, sequentially ordered by primary key, randomly linked ordered by primary key, and so forth.

3. It also results in space saving as compared with the other file structures.

Disadvantages of Inverted File Organization
Since the index entries are of variable lengths, index maintenance becomes more complex.

11.11   Summary

- A file is a collection of records. It is usually stored on a secondary storage device.

- The data is organized in a hierarchical order in the files. The hierarchical order includes items such as records, fields, and so on.

- File creation is the first operation to be performed on the files if the file is not created. A file is created by specifying its name and mode.

- A file is classified into two types, which are text files and binary files.

- A text file, often called a flat file, is a file that stores all the numeric or non-numeric data using its corresponding ASCII values. The data can be a string of letters, numbers, or special symbols.

- A binary file is a file that contains all the data in the binary form of 1s and 0s. It stores the data in the same form as that of primary memory.

- A file organization refers to the way in which records are physically arranged on a storage device.

- Sequence file organization is the most basic way to organize a collection of records in a file. In sequence file organization the file is created when the records are written, one after the other in order, and can be accessed only in the order in which they are written when the file is used for input. All the records are numbered from zero onward.

- An indexed sequential file organization is an effective way of organizing the records when there is a need to access both sequentially by some key values and also individually by the same key value. It provides the combination of access types that are supported by a sequential file or a relative file.

- Relative file organization provides an effective way of accessing individual records directly. In a relative file organization, there is a predictable relationship between the key and the record's location in the file.

- One fundamental approach for providing a linkage between an index and a file is called inversion. The inversion approach for providing multi-key access has been used as the basis for the physical data structure.

11.12   Exercises

11.12.1   Review Questions

1. What is a file?

2. Why is there a need to store the data in the files? Explain.

3. What do you understand about the terms record and field?

4. Discuss various operations that can be performed on files.

5. Differentiate between a text file and a binary file.

6. Write a short note on file attributes.

7. What do you understand about file organization? Discuss in detail.

8. Explain sequential file organization.

9. What are inverted files? Discuss.

10. Explain indexed sequential file organization.

11. Give the merits and drawbacks of indexed sequential file organization.

12. What is relative file organization? Also, discuss the advantages and disadvantages of relative file organization.

11.12.2 Multiple Choice Questions

1. A collection of related fields is called:
   (a) Data
   (b) Record
   (c) Field
   (d) File

2. A file marked as _____ can't be modified or deleted.
   (a) Hidden
   (b) Read-only
   (c) Archive
   (d) None of these

3. Which of the following is often known as a flat file?
   (a) Binary File
   (b) Text File
   (c) String File
   (d) None of these

4. _____ is a collection of data organized in a fashion which facilitates various operations such as updating, retrieving, and so forth.
   (a) Record
   (b) Data word
   (c) Field
   (d) File

5. Can relative files be used both for random as well as sequential access?
   (a) True
   (b) False
   (c) Not possible to comment

6. A file marked as _____ is not displayed in the directory.
   (a) Read-only
   (b) Archive
   (c) Hidden
   (d) None of these

7. A data field is characterized by:
   (a) Type
   (b) Size
   (c) Mode
   (d) Both (a) and (b)

8. _____ is used to store a collection of files.
   (a) Record
   (b) Dictionary
   (c) Directory
   (d) System

CHAPTER 12

GRAPHS

12.1 Introduction

So far, we have studied various types of linear data structures which are widely used in various applications. But the only non-linear data structure we have studied thus far is trees. In trees, we discussed the parent-child relationship in which one parent can have many children. But in graphs, this parent-child relationship is less restricted, that is, any complex relationship can exist. Thus, a tree can be generalized as a special type of graph. Therefore, a graph is a non-linear data structure which has a wide range of real-life applications. *A graph is a collection of some vertices (nodes) and edges that connect these vertices.* Figure 12.1 represents a graph.

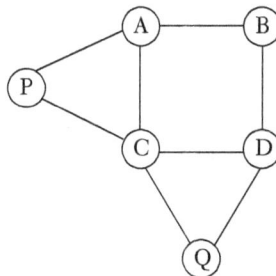

Figure 12.1. A graph.

Thus, *a graph G can be defined as an ordered set of vertices and edges (V, E), where V(G) represents the set of vertices and E(G) represents the set of edges that connect these vertices.* In the previous figure, V(G) = {A, B, C, D,

P, Q} represents the set of vertices, and E(G) = {(A, B), (B, D), (D, C), (C, A), (C, Q), (Q, D), (A, P), (P, C)} represents the set of edges.

Practical Application

A simple illustration of a graph is that when we connect with our friends on social media, say Facebook, where each user is a vertex and two users connect with each other, it forms an edge.

There are two types of graphs:

1. **Undirected Graph:** In an undirected graph, the edges do not have any direction associated with them. As we can see in the following figure, the two nodes A and B can be traversed in both directions, that is, from A to B or from B to A. Thus, an undirected graph does not give any information about the direction.

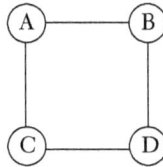

Figure 12.2. An undirected graph.

2. **Directed Graph:** In a directed graph, the edges have directions associated with them. As we can see in the following figure, the two nodes A and B can be traversed in only one direction, that is, only from A to B and not from B to A. Therefore, in the edge (A, B), the node A is known as the initial node and node B is known as the final node.

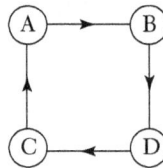

Figure 12.3. A directed graph.

12.2 Definitions

- **Degree of a vertex/node:** The degree of a node is the total number of edges incident to that particular node. Here, the degree of node B is three, as three edges are incident to node B.

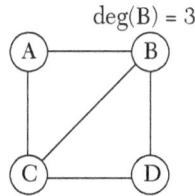

Figure 12.4. Graph showing degree of node B.

- **In-degree of a node:** The in-degree of a node is equal to the number of edges arriving at that particular node.

- **Out-degree of a node:** The out-degree is equal to the number of edges leaving that particular node.

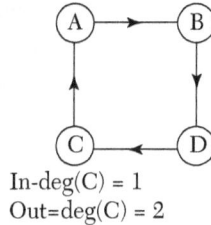

In-deg(C) = 1
Out=deg(C) = 2

Figure 12.5. Graph showing in-degree and out-degree of node C.

- **Isolated Node/Vertex:** A node having zero edges is known as an isolated node. The degree of such a node is zero.

Figure 12.6. Two isolated nodes X and Y.

- **Pendant Node/Vertex:** A node having one edge is known as a pendant node. The degree of such a node is one.

Figure 12.7. Two pendant nodes X and Y.

- **Adjacent Nodes:** For every edge e = (A, B) that connects nodes A andB, the nodes A and B are said to be adjacent nodes.

- **Parallel Edges:** If there is more than one edge between the same pair of nodes, then they are known as parallel edges.

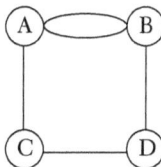

Figure 12.8. Parallel edges between A and B.

▪ **Loop:** If an edge hasa starting and ending point at the same node, that is, e = (A, A), then it is known as a loop.

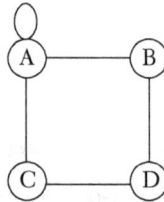

Figure 12.9. A loop.

▪ **Simple Graph:** A graph G(V, E) is known as a simple graph if it does not contain any loops or parallel edges.

▪ **Complete Graph:** A graph G(V, E) is known as a complete graph if and only if every node in the graph is connected to another node and there is no loop on any of the nodes.

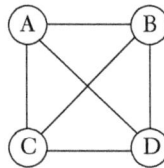

Figure 12.10. Complete graph.

▪ **Regular Graph:** A regular graph is a graph in which every node has the same degree. If every node has a degree r, then the graph is called a regular graph of degree r. In the given figure, all the nodes have the same degree, that is, 2; hence, it is known as a 2-regular graph.

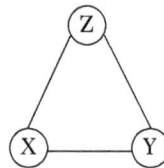

Figure 12.11. 2-Regular graph.

▪ **Multi-graph:** A graph G(V, E) is known as a multi-graph if it contains a loop, parallel edges, or both.

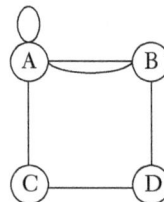

Figure 12.12. Multi-graph.

- **Cycle:** It is a path containing one or more edges which start from a particular node and also terminate at the same node.

- **Cyclic Graph:** A graph which has cycles in it is known as a cyclic graph.

- **Acyclic Graph:** A graph without any cycles is known as an acyclic graph.

- **Connected Graph:** A graph G(V, E) is known as a connected graph if there is a path from any node in the graph to another node in the graph such that for every pair of distinct nodes, there must be a path.

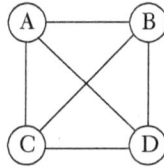

Figure 12.13. Connected graph.

- **Strongly Connected Graph:** A directed graph is said to be a strongly connected graph if there exists a dedicated path between every pair of nodes in the graph.For example, if there are two nodes, say P and Q, and there is a dedicated path from P to Q, then there must be a path from Q to P.

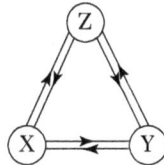

Figure 12.14. Strongly connected graph.

- **Size of a Graph:** The size of a graph is equal to the total number of edges present in the graph.

- **Weighted Graph:** A graph G(V, E) is said to be a weighted graph if all the edges in the graph are assigned some data. This data indicates the cost of traversing the edge.

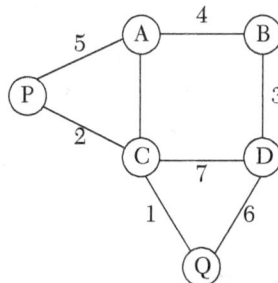

Figure 12.15. Weighted graph.

12.3 Graph Representation

Graphs can be represented in a computer's memory in either of the following ways:

1. Sequential Representation of Graphs using an Adjacency Matrix

2. Linked Representation of Graphs using an Adjacency List

Now, let us discuss both in detail.

12.3.1 Adjacency Matrix Representation

An adjacency matrix is used to represent the information of the nodes which are adjacent to one another. The two nodes will only be adjacent when there is an edge connecting those nodes. For any graph G having n nodes, the dimension of the adjacency matrix will be (n X n). Let G(V, E) be a graph having vertices V = {$V_1, V_2, V_3.........V_n$}, and then the adjacency matrix representation (n X n) will be given by:

$$a_{ij} = \begin{cases} 1 & \text{if there is an edge from } V_i \text{ to } V_j \\ 0 & \text{otherwise} \end{cases}$$

The adjacency matrix is also known as a bit matrix or Boolean matrix, since it contains only 0s and 1s. Now, let us take few examples to discuss and understand it more clearly.

Example 1: Consider the given directed graph and find its adjacency matrix.

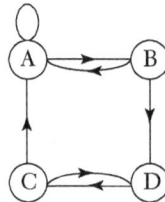

Figure 12.16. A directed graph.

The adjacency matrix for the graph will be:

$$\begin{array}{c} \quad A \ B \ C \ D \\ \begin{array}{c} A \\ B \\ C \\ D \end{array} \begin{bmatrix} 1 & 1 & 0 & 0 \\ 1 & 0 & 0 & 1 \\ 1 & 0 & 0 & 1 \\ 0 & 0 & 1 & 0 \end{bmatrix} \end{array}$$

Example 2: Now, consider the given undirected graph and find its adjacency matrix.

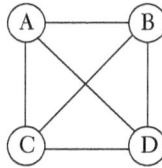

Figure 12.17. An undirected graph.

The adjacency matrix for the graph will be:

$$
\begin{array}{c c c c c}
 & A & B & C & D \\
\begin{matrix} A \\ B \\ C \\ D \end{matrix} &
\begin{bmatrix}
0 & 1 & 1 & 1 \\
1 & 0 & 1 & 1 \\
1 & 1 & 0 & 1 \\
1 & 1 & 1 & 0
\end{bmatrix}
\end{array}
$$

Example 3: Now, consider the given weighted graph and find its adjacency matrix.

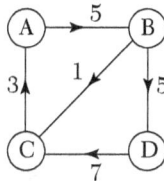

Figure 12.18. A directed weighted graph.

The adjacency matrix for the graph will be:

$$
\begin{array}{c c c c c}
 & A & B & C & D \\
\begin{matrix} A \\ B \\ C \\ D \end{matrix} &
\begin{bmatrix}
0 & 5 & 0 & 0 \\
0 & 0 & 0 & 4 \\
3 & 1 & 0 & 0 \\
0 & 0 & 7 & 0
\end{bmatrix}
\end{array}
$$

Example 4: Consider the given undirected multi-graph and find its adjacency matrix.

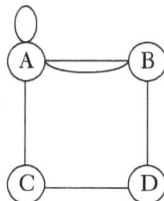

Figure 12.19. An undirected multi-graph.

The adjacency matrix for the graph will be:

$$\begin{array}{c}\begin{array}{cccc}A & B & C & D\end{array}\\\begin{array}{c}A\\B\\C\\D\end{array}\begin{bmatrix}1 & 2 & 1 & 0\\2 & 0 & 0 & 1\\1 & 0 & 0 & 1\\0 & 1 & 1 & 0\end{bmatrix}\end{array}$$

From the previous examples, we conclude that:

- The memory space needed to represent a graph using its adjacency matrix is n^2 bits.
- The adjacency matrix for an undirected graph is always symmetric.
- The adjacency matrix for a directed graph needs not be symmetric.
- The adjacency matrix for a simple graph having no loops or parallel edges will always contain 0s on the diagonal.
- The adjacency matrix for a weighted graph will always contain the weights of the edges connecting the nodes instead of 0 and 1.
- The adjacency matrix for an undirected multi-graph will contain the number of edges connecting the vertices instead of 1.

12.3.2 Adjacency List Representation

Adjacency matrix representation has some major drawbacks. First, it is very difficult to insert and delete the nodes in/from the graph as the size of the matrix needs to be changed accordingly, which is a very time-consuming process. Also, sometimes the matrix may contain many zeroes (sparse matrix). Hence, it is not a healthy representation. Therefore, adjacency list representation is preferred for representing sparse graphs in the memory. In this representation, every node is linkedto its list of all the other nodes which are adjacent to it.Adjacency list representation makes it easier to add or delete nodes. Also, it shows the adjacent nodes of a particular node. Now, let us take a few examples to discuss and understand it more clearly.

Example 1: Consider the given undirected graph and find its adjacency list representation.

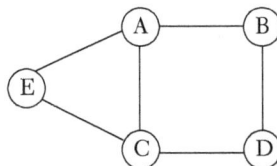

Figure 12.20. An undirected graph.

The adjacency list representation of the graph will be:

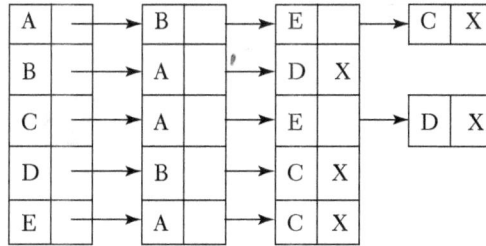

Example 2: Consider the given directed graph and find its adjacency list representation.

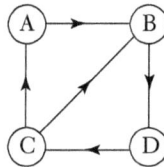

Figure 12.21. A directed graph.

The adjacency list representation of the graph will be:

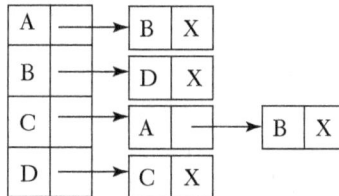

Example 3: Now, consider the given weighted graph and find its adjacency list representation.

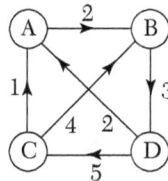

Figure 12.22. A directed weighted graph.

The adjacency list representation of the graph will be:

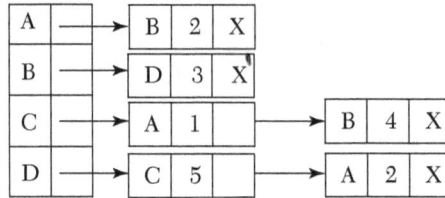

12.4 Graph Traversal Techniques

In this section, we will discuss various types of techniques to traverse a graph. As we all know, a graphis a collection of nodes and edges. Thus, traversing a graph is the process of visiting each node and edge in some systematic approach. Therefore, there are two types of standard graph traversal techniques, which are:

1. Breadth First Search (BFS)

2. Depth First Search (DFS)

Now we will discuss both these techniques in detail.

12.4.1 Breadth First Search

Breadth first search is a traversal technique that uses the queue as an auxiliary data structure for traversing all member nodes of a graph. In this technique, first we will select any node in the graph as a starting node, and then we will take all the nodes adjacent to the starting node. We will maintain the same approach for all the other nodes. Also, we will maintain the status of all the traversed/visited nodes in a queue so that no nodes are traversed again. Now, let us take a graph and apply BFS to traverse the graph.

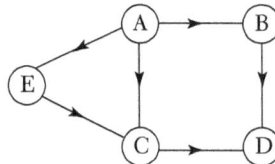

Figure 12.23. A sample graph.

Now, we will start the traversal of the graph by taking node A as a starting node of the previous sample graph. Then, we will traverse all the nodes adjacent to starting node A. As we can see, B, C, and E are the adjacent nodes of A. So, we will traverse these nodes in any order, say E, C, B. So, the traversal is:

$$\boxed{\text{A, E, C, B}}$$

Now, we will traverse all the nodes adjacent to E. Node C is adjacent to node E. But node C has already been traversed, so we will ignore it and we will move to the next step. Now, we will traverse all the nodes adjacent to node C. As we can see, D is the adjacent node of C. So, we will traverse node D and the traversal is:

$$\boxed{\text{A, E, C, B, D}}$$

Now we can see that all the nodes have been traversed, and hence this was the breadthfirst search traversal by taking node A as a starting node.

Now, we will implement the breadthfirst search traversal technique with the help of a queue. In this, we will maintain an array which will store all the adjacent unvisited neighbornodes of a given nodeunder consideration. Initially, the front and rear are set to -1. We will also maintain the status of the visited nodes in a Boolean array, which will have value 1if the node is visited and 0 if it is not visited.

▪ First, we willen-queue/insert the starting node into the queue.

▪ Second, the first node/element in the queue is deleted from the queue and all the adjacent unvisited nodes are inserted into the queue. This is repeated until the queue becomes empty.

For example: Consider the following sample graph and traverse the graph using the breadth first search technique.

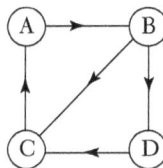

Figure 12.24. A sample graph.

The appropriate adjacency list representation of the graph is given as follows:

Node	Adjacency List
A	B, C
B	C
C	D
D	B

In this example, we are taking A as a starting node.

Step 1: First, node A is inserted into the queue.

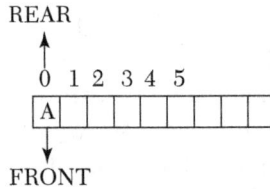

```
              REAR
               ↑
        0  1 2  3 4  5
       ┌─┬─┬─┬─┬─┬─┬─┐
       │A│ │ │ │ │ │ │
       └─┴─┴─┴─┴─┴─┴─┘
        ↓
       FRONT
```

Step 2: Node A is deleted from the queue and FRONT is incremented by 1. Now, insert all the nodes adjacent to A, which are nodes B and C, by incrementing REAR. Also, node A has been traversed.

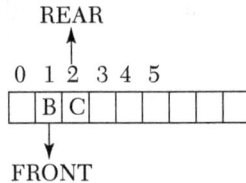

```
               REAR
                ↑
        0  1 2  3 4  5
       ┌─┬─┬─┬─┬─┬─┬─┐
       │ │B│C│ │ │ │ │
       └─┴─┴─┴─┴─┴─┴─┘
          ↓
        FRONT
```

Step 3: Similarly, node B is deleted from the queue and FRONT is incremented by 1. Now, insert all the nodes adjacent to B, which is node C, by incrementing REAR. But C has already been inserted in the queue. So now in this case, node C is also deleted by incrementing FRONT by 1, and the node adjacent to C, that is, D, is inserted into the queue. Therefore, nodes A, B, and C are traversed.

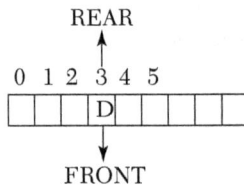

```
               REAR
                ↑
        0  1 2  3 4  5
       ┌─┬─┬─┬─┬─┬─┬─┐
       │ │ │ │D│ │ │ │
       └─┴─┴─┴─┴─┴─┴─┘
              ↓
            FRONT
```

Step 4: Now we will again delete the front element from the queue, which is D. We will insert the adjacent node of D, that is, B. But it is already traversed. Finally, as we delete the front element D, we notice that FRONT > REAR, which is not possible. Hence, we have traversed all the nodes in the graph.

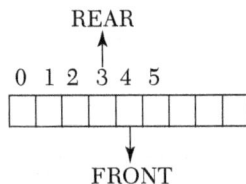

```
               REAR
                ↑
        0  1 2  3 4  5
       ┌─┬─┬─┬─┬─┬─┬─┐
       │ │ │ │ │ │ │ │
       └─┴─┴─┴─┴─┴─┴─┘
               ↓
             FRONT
```

Therefore, the breadthfirst search traversal of the graph is given as:

A, B, C, D

Now, let us look at the function for a breadth first search traversal.

//Write a program for breadth first search traversal

```java
import java.util.HashMap;
import java.util.LinkedList;
public class Graph {
    private intnumV;
    private boolean[][] matrix;
    public Graph(int v) {
        numV = v;
        matrix = new boolean[numV + 1][numV + 1];
    }
    public void addEdge(int u, int v) {
        matrix[u][v] = true;
        matrix[v][u] = true;
    }
    public boolean BFS(int source, int destination) {
        //Using predefined HashMap and LinkedList
        HashMap<Integer, Boolean> visited = new HashMap<>();
        LinkedList<Integer> queue = new LinkedList<>();
        queue.addLast(source);
        while (!queue.isEmpty()) {
            int r = queue.removeFirst();
            if (visited.containsKey(r)) {
                continue;
            }
            visited.put(r, true);
            if (r == destination) {
                return true;
            }
            for (int i = 1; i< matrix[0].length; i++) {
                if (matrix[r][i] && !visited.containsKey(i)) {
                    int n = i;
                    queue.addLast(n);
                }
            }
        }
        return false;
    }
}
//CLIENT CLASS
```

```java
import java.util.Scanner;
   public class GraphClient {
       public static void main(String[] args) {
       Scanner scn = new Scanner(System.in);
       Graph g = new Graph(7);
       g.addEdge(1, 2);
       g.addEdge(1, 4);
       g.addEdge(2, 3);
       g.addEdge(3, 4);
       g.addEdge(4, 5);
       g.addEdge(5, 6);
       g.addEdge(6, 7);
       g.addEdge(5, 7);
       int v1, v2;
       System.out.println("Enter the pair to be searched:");
       v1 = scn.nextInt();
       v2 = scn.nextInt();
       if (g.BFS(v1, v2))
           System.out.println("Pair found");
       else
           System.out.println("Pair not found");
   }
}
```

The output of the program is shown as:

12.4.2 Depth First Search

Depth first search is another traversal technique that uses the stack as an auxiliary data structure for traversing all the member nodes of a graph. Also in this technique, we first select any node in the graph as a starting node, and then we travel along a path which begins from the starting node. We will visit the adjacent node of the starting node, and again the adjacent node of the previous node, and so on. We will maintain the same approach for all the other nodes. Now, let us take a graph and apply DFS to traverse the graph.

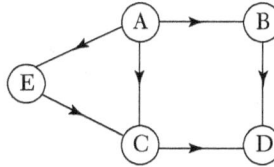

Figure 12.25. A sample graph.

Now we will start the traversal of the graph by taking node A as a starting node. Then we will traverse any of the nodes adjacent to the starting node A. As we can see, B, C, and E are the adjacent nodes of A. If we traverse node E, then we will traverse the node adjacent to E, that is, C. After traversing C, we will traverse the node adjacent to C, which is D. Now, there is no adjacent node to D; hence, we have reached the dead end. Thus, the traversal until now is:

A, E, C, D

Because of the dead end, we will move backward. Now, we reach node C. We will check if there is any other node adjacent to C. There is no such node, and thus we again move backward. Now, we reach E. We will again check if there is any other node adjacent to E. There is no such node, and thus we again move backward. Now, we reach A. We will check if there is any other node adjacent to A. There are two nodes, B and C, adjacent to node A. As C is already traversed, it will be ignored. Now, we will traverse node B. After traversing B, we will traverse the node adjacent to B, which is D, but D is already traversed. We can't move backward or forward. Thus, we have completed the traversal. The final traversal is given as:

A, E, C, D, B

Now we will implement the depth first search traversal technique with the help of a stack. In this, we will maintain an array which will store all the adjacent unvisited neighbor nodes of a given node. Initially, the top is set to -1. We will also maintain the status of the visited nodes in a Boolean array, which will have value 1 if the node is visited and 0 if it is not visited.

- First, we will push the starting node onto the stack.

- Second, the topmost node/element is popped out from the stack and is traversed. If it is already traversed, then we will ignore it.

- Third, all the adjacent unvisited nodes of the popped node/element are pushed onto the stack. This process is repeated until the queue becomes empty. The steps are repeated until the stack becomes empty.

For example: Consider the following sample graph and traverse the graph using the breadth first search technique.

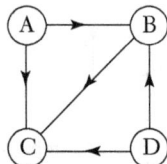

Figure 12.26. A sample graph.

In this example, we are taking A as a starting node.

Step 1: Push A onto the stack.

A

Step 2: Now, pop the topmost element from the stack, that is, A. Thus, A is traversed. Now, push all the nodes adjacent to A, that is, push B and C.

B, C

Step 3: Again, pop the topmost element from the stack, that is, C. Thus, C is also traversed. Now, push all the nodes adjacent to C, that is, push D.

B, D

Step 4: Now, again pop the topmost element from the stack, that is, D. Thus, D is also traversed. Now, push all the nodes adjacent to D, that is, push B. But B is already in the stack. Therefore, no push is performed. Thus, the stack becomes:

B

Step 5: Again, pop the topmost element from the stack, that is, B. Thus, B is also traversed. Now, push all the nodes adjacent to B, that is, push C. But C is already traversed; hence, the stack becomes empty.

Therefore, the depthfirst search traversal of the graph is given as follows:

A, C, D, B

Now, let us look at the function for the depth first search traversal.

//Write a program for depth first search traversal

```java
import java.util.HashMap;
import java.util.LinkedList;
public class Graph {
    private intnumV;
    private boolean[][] matrix;
    public Graph(int v) {
        numV = v;
        matrix = new boolean[numV + 1][numV + 1];
    }
    public void addEdge(int u, int v) {
        matrix[u][v] = true;
        matrix[v][u] = true;
    }
    public boolean DFS(int source, int destination) {
        //Using predefined HashMap and LinkedList
        HashMap<Integer, Boolean> visited = new HashMap<>();
        LinkedList<Integer> stack = new LinkedList<>();
        stack.addFirst(source);
        while (!stack.isEmpty()) {
            int r = stack.removeFirst();
            if (visited.containsKey(r))
                continue;
            visited.put(r, true);
            if (r == destination)
                return true;
            for (inti = 1; i< matrix[0].length; i++) {
                if (matrix[r][i] && !visited.containsKey(i)) {
                    int n = i;
                    stack.addFirst(n);
                }
            }
        }
        return false;
    }
}
//CLIENT CLASS
import java.util.Scanner;
public class GraphClient {
    public static void main(String[] args) {
        Scanner scn = new Scanner(System.in);
        Graph g = new Graph(7);
        g.addEdge(1, 2);
```

```
    g.addEdge(1, 4);
    g.addEdge(2, 3);
    g.addEdge(3, 4);
    g.addEdge(5, 6);
    g.addEdge(6, 7);
    g.addEdge(7, 8);
    int v1, v2;
    System.out.println("Enter the pair to be searched:");
    v1 = scn.nextInt();
    v2 = scn.nextInt();
    if (g.DFS(v1, v2))
        System.out.println("Pair found");
    else
        System.out.println("Pair not found");
    }
}
```
The output of the program is shown as:

Memory Aid

To remember which of the data structuresare used in implementing a breadth first search and a depthfirst search, we can use this memory aid. Breadth first search is implemented using a queue data structure, and depth first search is implemented using a stack data structure, as can be remembered by alphabetical order. B (Breadth First Search) and Q (Queue) comes before than D (Depth First Search) and S (Stack) in alphabetical order.

12.5 Topological Sort

Topological sort is a procedure to determine the linear ordering of the nodes of an acyclic directed graph(also known as DAG), in which each node comes before all those nodes which have zero predecessors. A topological sort of a DAG is a linear ordering of the vertices of a graph G(V, E) such that if(a, b) is an edge, then a must appear before b in the topological ordering. The main idea behind this is that in a graph, if a vertex has in-degree 0, then that vertex should be selected as the first element in the topological order. Also, a topological sort is possible only in acyclic directed graphs. An acyclic graph is one which does not have any cycles in it. Topological sorting is widely used in scheduling tasks, applications, and so on. Now, let us look at the algorithm of topological sorting.

Algorithm for Topological Sort

```
Step 1: START
Step 2: Find the in-degree of every node.
Step 3: Insert all the nodes/elements having in-degree zero in the
        queue.
Step 4: Repeat Steps 5 and 6 until the queue becomes empty.
Step 5: Delete the first node from the queue by incrementing FRONT
        by 1.
Step 6: Repeat for each neighbor P of node N -
        (a) Delete the edge from P to M by decreasing the in-de-
            gree by 1.
        (b) If in-degree of P is zero, then add P to the rear of
            the queue.
Step 7: END
```

For example: Consider a given acyclic directed graph and find its topological sort.

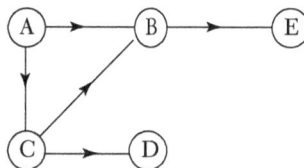

Figure 12.27. Acyclic directed graph.

The appropriate adjacency list representation of the previous graph is given as follows:

Node	Adjacency List
A	B, C
B	E
C	B, D
D	-
E	-

Step 1: In-degree of all the nodes:

    In-degree (A) – 0

    In-degree (B) – 2

    In-degree (C) – 1

    In-degree (D) – 1

    In-degree (E) – 1

Now, we have node A with in-degree = 0; thus, A will be added to the queue.

Step 2: Now, insert node A into the queue.

FRONT = 1, REAR = 1, QUEUE = A

Step 3: Now, delete node A from the queue. Also, delete all the edges going from A.

FRONT = 0, REAR = 0, TOPOLOGICAL SORT = A

Thus, the graph becomes:

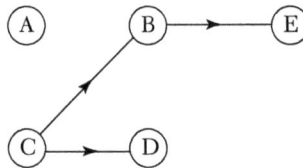

Now, the in-degree of all the nodes:

    In-degree (B) – 1

    In-degree (C) – 0

    In-degree (D) – 1

    In-degree (E) – 1

■ Now, we have node C with in-degree = 0; thus, C will be added to the queue.

Step 4: Now, insert node C into the queue.

FRONT = 1, REAR = 1, QUEUE = C

Step 5: Now, delete node C from the queue. Also, delete all the edges going from C.

FRONT = 0, REAR = 0, TOPOLOGICAL SORT = A, C

Thus, the graph becomes:

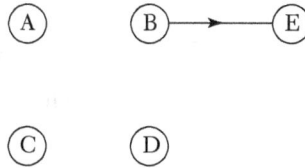

Now, the in-degree of all the nodes:

In-degree (B) – 0

In-degree (D) – 0

In-degree (E) – 1

Now, we have two nodes B and D with in-degree = 0; thus, B and D will be added to the queue.

Step 6: Now, insert nodes B and D into the queue.

FRONT = 1, REAR = 2, QUEUE = B, D

Step 7: Now, delete node B from the queue. Also, delete all the edges going from B. There will be no change in the in-degree of the nodes.

FRONT = 1, REAR = 1,
TOPOLOGICAL SORT = A, C, B, QUEUE = D

Step 8: Now, delete node D from the queue. Also, delete all the edges going from D.

FRONT = 0, REAR = 0, TOPOLOGICAL SORT = A, C, B, D

Thus, the graph becomes:

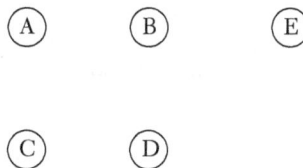

Now, the in-degree of all the nodes:

In-degree (E) – 0

Now, we have node E with in-degree = 0. Thus, E will be added to the queue.

Step 9: Now, insert node E into the queue.

FRONT = 1, REAR = 1, QUEUE = E

Step 10: Now, delete node E from the queue. Also, delete all the edges going from E.

FRONT = 0, REAR = 0, TOPOLOGICAL SORT = A, C, B, D, E

Now, we have no nodes left in the graph. Thus, the topological sort of the graph will be

> A, C, B, D, E

12.6 Minimum Spanning Tree

A spanning tree of an undirected and connected graph G is a sub-graph which contains all the vertices and edges that connect these vertices and is a tree. The weights/costs can be assigned to the edges, and these weights/costs can be used to calculate the weight/cost of the spanning tree by calculating the sum of the weights/costs of each edge. A graph can have many spanning trees. Thus, a minimum spanning tree (MST) is defined as a spanning tree that has weights/costs associated with the edges such that the total weight/cost of the spanning tree is at a minimum. Although there are various approaches for determining an MST, the two most popular approaches for determining a minimum cost spanning tree of a graph are:

1. Prim's Algorithm

2. Kruskal's Algorithm

Now, let us discuss both of them in detail.

12.6.1 Prim's Algorithm

Prim's algorithm is the algorithm that is used to build a minimum cost spanning tree. This algorithm works in such a way that it builds a tree edge by edge. The next edge to be included is chosen according to some criterion. The steps involved in Prim's algorithm are:

Step 1: Select a starting vertex/node and add it to the spanning tree.

Step 2: During each iteration, select a vertex/node in such a way that the edge connecting vertex V_i to another vertex V_j has the minimum cost/weight assigned to it. Remember, the edge forming a cycle must not be added.

Step 3: End the process when (n-1) number of edges have been inserted into the tree.

Frequently Asked Questions

1. Consider the given graph and construct a minimum spanning tree using Prim's algorithm.

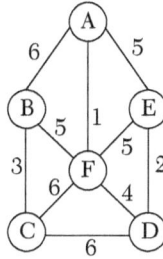

Ans:

Step 1: *The starting node is F.*

Step 2: *The lowest weighted/cost edge is (F, A), that is, 1. Hence, it is added to the tree.*

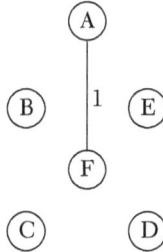

Step 3: *Now, the lowest weighted/cost edge is (F, D), that is, 4. Hence, it is added to the tree.*

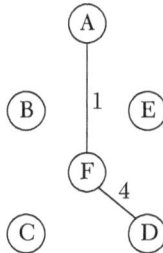

Step 4:

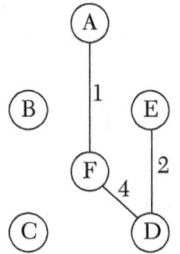

Step 5:

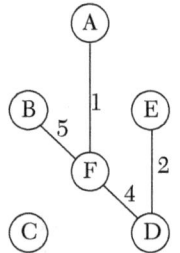

Step 6: *Finally, the minimum spanning tree is constructed.*

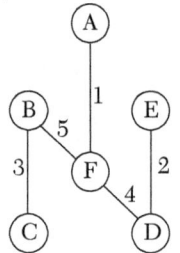

12.6.2 Kruskal's Algorithm

Kruskal's algorithm is another approach for determining the minimum cost spanning tree of a graph. In this approach also, the tree is built edge by edge. The next edge to be included is chosen according to some criterion. The steps involved in Kruskal's algorithm are:

Step 1: The weights/costs assigned to the edges are sorted in ascending order.

Step 2: In this step, the lowest weighted/cost edge is added to the tree. Remember, the edge forming a cycle must not be added.

Step 3: End the process when (n-1) number of edges have been inserted into the tree.

Frequently Asked Questions

2. Consider the given graph and construct a minimum spanning tree using Kruskal's algorithm.

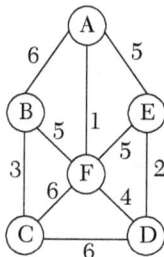

Ans:

Step 1: *Initially the tree is given as:*

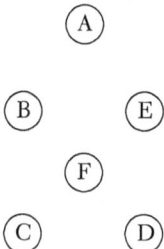

Step 2: *Choose edge (F, A).*

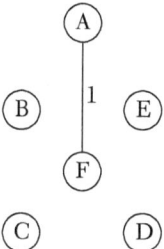

Step 3: *Choose edge (D, E).*

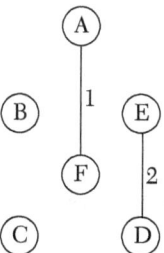

Step 4: *Choose edge (B, C).*

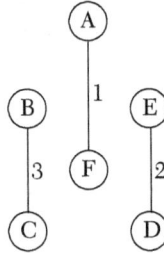

Step 5: *Choose edge (F, D).*

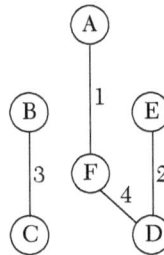

Step 6: *Choose edge (F, B).*

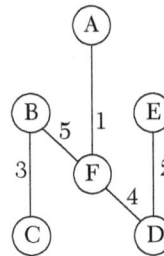

Practical Application

Graphs are used to find the shortest route using GPS, Google Maps, and Yahoo!Maps.

12.7 Summary

- A graph is a collection of vertices (nodes) and edges that connect these vertices.

- The degree of a node is the total number of edges incident to that particular node.

- A graph G (V, E) is known as a complete graph if and only if every node in the graph is connected to another node and there is no loop on any of the nodes.

- An adjacency matrix is usually used to represent the information of the nodes which are adjacent to one another. The adjacency matrix is also known as a bit matrix or Boolean matrix since it contains only 0s and 1s.

- In adjacency list representation, every node is linked to its list of all the other nodes which are adjacent to it.

- Traversing a graph is the process of visiting each node and edge in some systematic approach.

- Breadth first search is a traversal technique that uses the queue as an auxiliary data structure for traversing all the member nodes of the graph. In this technique, first we will select any node in the graph as a starting node, and then we will take all the nodes adjacent to the starting node. We will maintain the same approach for all the other nodes.

- Depth first search is another traversal technique that uses the stack as an auxiliary data structure for traversing all the member nodes of the graph. In this also, first we will select any node in the graph as a starting node, and then we will travel along a path which begins from the starting node. We will visit the adjacent node of the starting node, and again the adjacent node of the previous node, and so on.

- Topological sort is a procedure to determine linear ordering of the nodes of an acyclic directed graph (also known as DAG), in which each node comes before all those nodes which have zero predecessors.

- A minimum spanning tree (MST) is defined as a spanning tree that has weights/costs associated with the edges such that the total weight/cost of the spanning tree is at a minimum.

12.8   Exercises

12.8.1   Theory Questions

1. What is a graph? Explain its features.

2. What do you understand about a complete graph?

3. What is a multi-graph?

4. How can a graph be represented in the computer's memory? Discuss.

5. Differentiate between a directed and undirected graph with an example of each.

6. Consider the following graph and find the following:
   (a) Adjacency Matrix Representation.
   (b) Degree of each node.
   (c) Is the graph complete?
   (d) Pendant nodes.

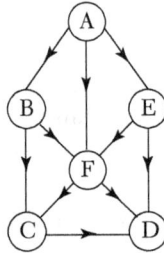

7. Explain why adjacency list representation is preferred for storing sparse matrices over adjacency matrix representation.

8. What are the different types of graph traversal techniques? Explain each of them in detail with the help of an example.

9. What do you understand about topological sort?

10. In what kind of graphs can topological sorting be used?

11. Differentiate between breadth first search and depth first search.

12. Consider the following graph and find out its BFS and DFS traversal.

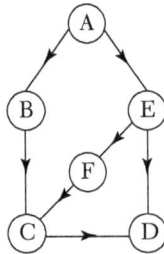

13. What is a spanning tree?

14. Why is a minimum spanning tree called a spanning tree? Discuss.

15. Consider the given adjacency matrix and draw the directed graph.

$$
\begin{array}{c@{\quad}cccc}
 & A & B & C & D \\
A & \begin{bmatrix} 1 \\ 1 \\ 0 \\ 0 \end{bmatrix} & \begin{matrix} 1 \\ 1 \\ 1 \\ 1 \end{matrix} & \begin{matrix} 0 \\ 1 \\ 1 \\ 1 \end{matrix} & \begin{matrix} 1 \\ 0 \\ 0 \\ 1 \end{bmatrix}
\end{array}
$$

$$
\begin{array}{cc}
 & \begin{array}{cccc} A & B & C & D \end{array} \\
\begin{array}{c} A \\ B \\ C \\ D \end{array} & \left[\begin{array}{cccc} 1 & 1 & 0 & 1 \\ 1 & 1 & 1 & 0 \\ 0 & 1 & 1 & 0 \\ 0 & 1 & 1 & 1 \end{array}\right]
\end{array}
$$

16. Write a short note on Prim's algorithm.

17. Explain Kruskal's algorithm.

18. List some of the real-life applications of graphs.

19. Consider the following graph and find the minimum spanning tree using
  (a) Prim's algorithm
  (b) Kruskal's algorithm

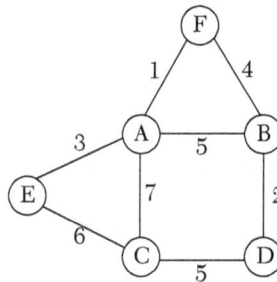

12.8.2 Programming Questions

1. Write a Java program to create and display a graph.

2. Write an algorithm to perform a topological sort on a graph.

3. Write an algorithm to find the degree of a node N in a graph.

4. Write a Java program to traverse a graph using depth first search.

5. Write an algorithm to traverse a graph using breadth first search.

6. Write a Java program to find the shortest path using Prim's algorithm.

7. Write a Java program to find the shortest path using Kruskal's algorithm.

12.8.3 Multiple Choice Questions

1. To implement a breadth first search, the data structure used is:
   (a) Stack
   (b) Queue
   (c) Trees
   (d) Linked List

2. A graph having multiple edges is known as a _____.
   (a) Connected Graph
   (b) Complete Graph
   (c) Simple Graph
   (d) Multi-graph

3. An edge having initial and end points at the same node is called:
   (a) Degree
   (b) Cycle
   (c) Loop
   (d) Parallel Edge

4. An adjacency matrix is also known as a:
   (a) Bit Matrix
   (b) Boolean Matrix
   (c) Both of the above
   (d) None of the above

5. To implement a depth first search, the data structure used is:
   (a) Stack
   (b) Queue
   (c) Trees
   (d) Linked List

6. Topological Sort is performed only on:
   (a) Cyclic Directed Graphs
   (b) Acyclic Directed Graphs
   (c) Both of the above
   (d) None of the above

7. Which one of the following nodes has a zero degree?

   (a) Simple node

   (b) Isolated node

   (c) Pendant node

   (d) None of the above

8. _____ is the total number of nodes in a graph.

   (a) Degree

   (b) In-degree

   (c) Out-degree

   (d) Size

9. A graph G can have many spanning trees.

   (a) True

   (b) False

   (c) Not possible to comment

Answers to Multiple Choice Questions

Chapter 1   Introduction to Data Structures

Multiple Choice Questions

1. (b)	2. (b)	3. (d)	4. (a)	5. (d)	6. (d)	7. (c)	8. (d)	9. (c)	10. (c)
11. (a)	12. (c)	13. (a)	14. (c)	15. (d)					

Chapter 2   Introduction to the Java Language

Multiple Choice Questions

1. (b)	2. (b)	3. (a)	4. (c)	5. (d)	6. (b)	7. (b)	8. (c)	9. (d)

Chapter 3   Arrays

Multiple Choice Questions

1. (c)	2. (b)	3. (b)	4. (c)	5. (a)	6. (c)	7. (d)	8. (a)	9. (b)	10. (a)
11. (a)									

Chapter 4   Linked Lists

Multiple Choice Questions

1. (b)	2. (d)	3. (d)	4. (b)	5. (d)	6. (a)	7. (c)	8. (c)	9. (b)

Chapter 5   Queues

Multiple Choice Questions

1. (a)	**2.** (a)	**3.** (b)	**4.** (a)	**5.** (b)	**6.** (c)	**7.** (c)	**8.** (b)	**9.** (b)	**10.** (a)

Chapter 6   Searching and Sorting

Multiple Choice Questions

1. (a)	**2.** (d)	**3.** (c)	**4.** (a)	**5.** (d)	**6.** (b)	**7.** (c)	**8.** (b)	**9.** (b)

Chapter 7   Stacks

Multiple Choice Questions

1. (b)	**2.** (b)	**3.** (c)	**4.** (a)	**5.** (b)	**6.** (a)	**7.** (c)	**8.** (d)	**9.** (c)

Chapter 8   Trees

Multiple Choice Questions

1. (d)	**2.** (c)	**3.** (c)	**4.** (d)	**5.** (d)	**6.** (d)	**7.** (c)	**8.** (b)	**9.** (a)	**10.** (d)
11. (b)	**12.** (c)	**13.** (b)	**14.** (c)	**15.** (d)					

Chapter 9   Multi-Way Search Trees

Multiple Choice Questions

1. (b)	**2.** (a)	**3.** (d)	**4.** (c)	**5.** (c)	**6.** (b)	**7.** (b)	**8.** (b)	**9.** (c)

Chapter 10   Hashing

Multiple Choice Questions

1. (c)	**2.** (a)	**3.** (d)	**4.** (b)	**5.** (a)	**6.** (c)	**7.** (b)	**8.** (b)	**9.** (a)

Chapter 11   Files

Multiple Choice Questions

1. (b)	**2.** (b)	**3.** (b)	**4.** (a)	**5.** (b)	**6.** (c)	**7.** (d)	**8.** (c)

Chapter 12   Graphs

Multiple Choice Questions

1. (b)	**2.** (a)	**3.** (c)	**4.** (c)	**5.** (a)	**6.** (b)	**7.** (b)	**8.** (a)	**9.** (a)

Index

A

Abstraction, 27
Arithmetic Operators, 32
Array(s), 4, 61–102
  2-D (Two-dimensional), 83–90
    Declaration of, 84–86
    Operations on, 86–90
  Applications of, 93
  Calculating the address of, 64–65
    3-D, 91–92
  Declaration, 62
  Definition, 61–62
  Initialization, 63–64
  Introduction, 61
  Multidimensional/N-Dimensional, 90–91
  Operations on, 65–83
    Deleting an element in, 71–76
    Inserting an element in, 66–71
    Merging of two arrays, 78–81
    Searching an element in, 76–78
    Sorting an array, 81–83
    Traversing, 65–66
  Sparse matrices, 93–96
    Representation of, 95–96
    Types of, 94
  Summary, 96–98
Assignment operators, 34

B

Binary tree, 8, 287, 291
Bitwise operators, 35
Bottom-up approach, 12

C

Class, 24
Column major order, 85
Conditional operators, 35

D

Data, 1
Data hiding, 24
Data structure(s), 1–20
  Abstract data types, 14–15
  Algorithm(s), 10–11
    Analyzing an, 13–14
    Approaches for designing an, 12–13
    Categories of, 13–14
    Developing an, 11
    Features and characteristics of, 11
  Application areas, 2
  Big O notation, 15
  Classification of, 4
  Exercises, 17
  Introduction, 1
  Operations on, 10

Summary, 16
Types of, 2–9
    Arrays, 4
    Graphs, 9
    Homogeneous and
        non-homogeneous, 3
    Linear and non-linear, 3
    Linked lists, 7
    Primitive and non-primitive, 3
    Queues, 5
    Stacks, 6
    Static and dynamic, 3
    Trees, 8
Do-while Loop, 47–49
Dynamic binding or dispatch, 28

E

Encapsulation, 24, 55

F

FIFO (First In First Out), 5, 163
File(s), 381–391
    C vs C++ vs Java File Handling, 384
    Classification of, 383
    Operations of, 382–383
        Creation and retrieving, 382
    Indexed sequential file organization,
        386–387
    Introduction, 382
    Inverted file organization, 388
    Organization of, 384
    Relative file organization, 387–388
        Address calculation technique, 387
        Directory lookup technique, 387
    Sequence file organization, 385–386
        Advantages and disadvantages, 386
    Summary, 388
    Terminologies, 381–382
For Loop, 49–51
Frank, Ed, 21

G

Gosling, James, 21
Graph(s), 393–424
    Definitions, 394–397
    Introduction, 393–394
        Undirected and directed, 394
    Minimum spanning tree, 414–418

Prim's algorithm, 414–416
Kruskal's algorithm, 416–418
Representation, 398–402
    Adjacency matrix representation,
        398–400
    Adjacency list representation, 400–402
Summary, 418
Topological sort, 410–414
Traversal techniques, 402–410
    Breadth First Search (BFS), 402–406
    Depth First Search (DFS), 406–410

H

Hashing, 353–380
    Double, 374–376
    Introduction, 353–356
        Difference between hashing and direct
            addressing, 354–355
        Hash functions, 356
        Hash tables, 355
    Functions, 356–376
        Chaining method, 358–363
        Collision, 358–376
        Division method, 356–357
        Folding method, 357–358
        Mid–square method, 357
        Open addressing method, 363–376
    Summary, 376–378

I

IF statement, 36–38
IF–ELSE statement, 38–40
Inheritance, 25
    Single, multilevel, and multiple, 25–26

J

Java language, 21–60
    Break and continue statements, 51–53
    Characteristics of, 22
    Character set used in, 28
    Compiling the program, 23–25
    Data types in, 30–31
    Decision control statements in, 36–45
    Introduction, 21
        Primary goals in the creation of, 21
    Looping or iterative statements in, 45–51
    Methods in, 54
        Library and user–defined, 54

Operators in, 32–36
Overview, 22–23
Object–Oriented programming (OOP),
    24–28
Structure of a, 31–32
Summary, 55–57
Tokens, 29–30
    Constants, 30
    Identifiers, 29
    Keywords, 29
    Variables, 30
Java Virtual Machine (JVM), 22

L

Last In First Out (LIFO), 5, 241
Linear probing, 363–368
Linear search, 76
Linked list(s), 7, 103–162
    Applications of, 159
    Definition of a, 103–105
    Introduction, 103
    Memory allocation in a, 105–106
    Operations on a circular, 123–133
        Deleting a new node in, 126–133
        Inserting a new node in, 123–125
    Operations on a doubly, 134–148
        Deleting a new node in, 139–148
        Inserting a new node in, 134–139
    Operations on a single, 106–122
        Concatenation of two, 117
        Deleting a node from, 112–116
        Inserting a new node in, 108–112
        Reversing, 117
        Searching for a given value, 107–108
        Sorting, 117
        Traversing a, 107
    Polynomial representation, 159
    Summary, 159
    Types of, 106–158
        Circular linked list, 122–133
        Doubly linked list, 133–148
        Header linked list, 149–158
        Singly linked list, 106–122
Logical Operators, 34–34

M

Members, 1
Message passing, 28

Modularization, 12
Multi-way search tree(s), 339–352
    B+ trees, 349
    B-trees, 340–350
        Applications of, 349
        Deletion in, 343–349
        Insertion in, 341–343
        Operations, 341–350
    Introduction, 339–340
    Summary, 350

N

Naughton, Patrick, 21
Nested IF-ELSE statement, 40–42
Nodes, 7, 283
NULL, 7

O

Object, 24

P

Polymorph, 26
Polymorphism, 26
    Types of, 27
Program, 11

Q

Quadratic probing, 368–374
    Advantages and disadvantages
        of, 373–374
Queue(s), 5, 163–200
    Applications of, 197
    Definition of, 163–164
    Implementation of, 164–170, 186–
        Using arrays, 164, 186–192
        Using linked lists, 164–170, 186–187
    Introduction, 163
    Operations on, 170–175
        Deletion, 171–175
        Insertion, 170–171
    Summary, 197
    Types of, 175–196
        Circular, 175–184
        De-queue (double-ended queue),
            192–196
        Priority, 185–192

R

Relational operators, 34
Row major order, 85

S

Searching, 201–214
Introduction, 201
   Binary search, 206–210
      Complexity of, 208
      Drawbacks of, 208–210
      Search algorithm, 206–208
   Interpolation search, 210–214
      Complexity of, 212–214
      Working of algorithm, 211
   Linear search or sequential search,
      201–205
      Drawback, 204–205
      Search algorithm, 203
Selection sort technique, 81–82
Sheridan, Mike, 21
Sorting, 214–239
   External, 214
   Summary, 236
   Types of, 215–235
      Bubble, 226–230
      Insertion, 218–221
      Merge, 221–226
      Quick or partition exchange, 230–235
      Selection, 215–218
Space complexity, 13
Stack(s), 241–281
   Applications of, 255–277
      Conversion from Infix Expression to
         Postfix Expression, 256–261
      Conversion from Infix Expression to
         Prefix Expression, 262–266
      Evaluation of Postfix Expression,
         266–270
      Evaluation of Prefix Expression,
         270–274
      Parenthesis Balancing, 270–277
      Polish and Reverse Polish Notations,
         255–256
   Definition of, 242
   Implementation of, 249–255
      Using arrays, 249
      Using linked lists, 249–255
   Introduction, 241

   Operations on, 243–249
      Peek, 245–249
      Pop, 244–245
      Push, 243–244
   Overflow and underflow in, 242–243
   Summary, 278
Switch statement, 42–45
Subscript, 5

T

Time complexity, 13
Time-space trade-off, 14
Top-down approach, 12
Tree(s), 283–338
   AVL (Adelson-Velski and Landis), 319–330
      Need of height-balanced, 319–320
      Operations on an, 320–321
      Rotations, 321–330
   Binary, 287–290
      Array and linked representation of, 289
      Complete, 288
      Extended, 288
      Memory representation of, 289
   Binary search, 291–305
      Deleting a node/key from, 297–303
      Deleting the entire, 303
      Determining the height of, 305
      Finding the largest node, 304– 305
      Finding the mirror image, 303–304
      Finding the smallest node, 304
      Inserting a node/key in, 295–297
      Searching for a node/key, 292–295
   Binary tree traversal methods, 305–319
      Creating a, 315–319
      In-order, 307–308
      Post-order, 308–315
      Pre-order, 306–307
   Definitions, 284–287
   Introduction, 283–284
   Summary, 330–338

U

Unary operators, 35

W

Warth, Chris, 21
WHILE Loop, 46–47

www.ingramcontent.com/pod-product-compliance
Lightning Source LLC
Chambersburg PA
CBHW080134220326
41598CB00032B/5068